Introduction to Genetics

Introduction to Genetics

Catherine Hastings

Larsen & Keller
www.larsen-keller.com

Introduction to Genetics
Catherine Hastings
ISBN: 978-1-64172-614-6 (Hardback)

Larsen & Keller

Published by Larsen and Keller Education,
5 Penn Plaza,
19th Floor,
New York, NY 10001, USA

Cataloging-in-Publication Data

Introduction to genetics / Catherine Hastings.
p. cm.
Includes bibliographical references and index.
ISBN 978-1-64172-614-6
1. Genetics. 2. Adaptation (Biology). 3. Biology. I. Hastings, Catherine.
QH430 .I58 2022
576.5--dc23

For more information regarding Larsen and Keller Education and its products, please visit the publisher's website www.larsen-keller.com

TABLE OF CONTENTS

PREFACE

This book is a culmination of my many years of practice in this field. I attribute the success of this book to my support group. I would like to thank my parents who have showered me with unconditional love and support and my peers and professors for their constant guidance.

Genetics is the branch of biology that focuses on the study of genetic variation, genes and heredity in organisms. Some of the major areas of study within this field are trait inheritance and molecular inheritance mechanisms of genes. It also studies the function and behavior of genes. The major sub-fields of genetics include epigenetics, molecular genetics and population genetics. Epigenetics focuses on the study of the heritable phenotype changes that do not involve alterations in the DNA sequence. Molecular genetics studies the function and structure of genes in organisms using genetic screens. Population genetics deals with the genetic differences present within and between populations. This textbook is compiled in such a manner, that it will provide in-depth knowledge about the theory and concepts of genetics. While understanding the long-term perspectives of the topics, it makes an effort in highlighting their impact as a modern tool for the growth of the discipline. This book is appropriate for those seeking detailed information in this area.

The details of chapters are provided below for a progressive learning:

Chapter - Introduction

Genes are the sequences of nucleotides in DNA or RNA that encode the synthesis of a gene product. The branch of biology that studies genes, heredity and genetic variations in organisms is referred to as genetics. All the diverse aspects of genes and genetics have been briefly introduced in this chapter.

Chapter - Branches of Genetics

There are numerous branches of genetics such as human genetics, behavior genetics, cytogenetics, molecular genetics, population genetics, plant genetics, microbial genetics and classical genetics. The chapter closely examines these branches of genetics to provide an extensive understanding of the subject.

Chapter - Significant Aspects of Genetics

Some of the fundamental components which are studied in genetics are genomes, nucleic acids, genotypes, phenotypes, alleles, etc. Major concepts that are studied within this field include genetic variation, genome sequencing and gene expression. The topics elaborated in this chapter will help in gaining a better perspective about these components and concepts of genetics.

Chapter - Genetic Modification and Engineering

The direct manipulation of the genes of an organism using biotechnology is referred to as genetic engineering. It comprises of a group of technologies to change the genetic construction of cells. The topics elaborated in this chapter will help in gaining a better perspective about genetic engineering as well as various genetically modified organisms such as animals and crops.

Chapter - Heredity and Evolution

The passing on of traits from parents to their offspring through sexual or asexual reproduction is known as heredity. The change in the heritable characteristics of biological populations over successive generations is known as evolution. The chapter closely examines the key concepts of heredity and evolution such as Mendelian inheritance, speciation and common descent to provide an extensive understanding of the subject.

Catherine Hastings

Introduction

CHAPTER 1

Genes are the sequences of nucleotides in DNA or RNA that encode the synthesis of a gene product. The branch of biology that studies genes, heredity and genetic variations in organisms is referred to as genetics. All the diverse aspects of genes and genetics have been briefly introduced in this chapter.

GENE

Gene is the unit of hereditary information that occupies a fixed position (locus) on a chromosome. Genes achieve their effects by directing the synthesis of proteins.

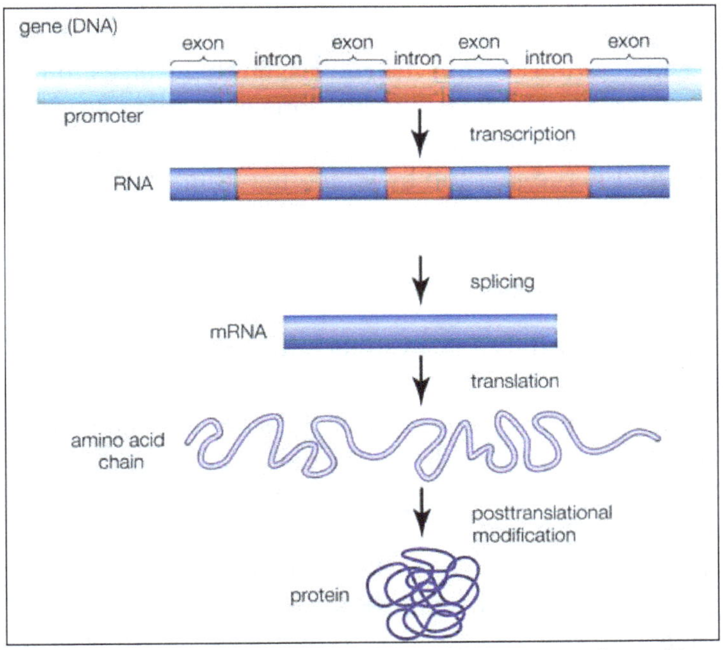

Genes are made up of promoter regions and alternating regions of introns (noncoding sequences) and exons (coding sequences).

The production of a functional protein involves the transcription of the gene from DNA into RNA, the removal of introns and splicing together of exons, the translation of the spliced RNA sequences into a chain of amino acids, and the posttranslational modification of the protein molecule.

In eukaryotes (such as animals, plants, and fungi), genes are contained within the cell nucleus. The mitochondria (in animals) and the chloroplasts (in plants) also contain small subsets of genes distinct from the genes found in the nucleus. In prokaryotes (organisms lacking a distinct nucleus, such as bacteria), genes are contained in a single chromosome that is free-floating in the cell cytoplasm. Many bacteria also contain plasmids—extrachromosomal genetic elements with a small number of genes.

The number of genes in an organism's genome (the entire set of chromosomes) varies significantly between species. For example, whereas the human genome contains an estimated 20,000 to 25,000 genes, the genome of the bacterium Escherichia coli O157:H7 houses precisely 5,416 genes. Arabidopsis thaliana—the first plant for which a complete genomic sequence was recovered—has roughly 25,500 genes; its genome is one of the smallest known to plants. Among extant independently replicating organisms, the bacterium Mycoplasma genitalium has the fewest number of genes, just 517.

Chemical Structure of Genes

Genes are composed of deoxyribonucleic acid (DNA), except in some viruses, which have genes consisting of a closely related compound called ribonucleic acid (RNA). A DNA molecule is composed of two chains of nucleotides that wind about each other to resemble a twisted ladder. The sides of the ladder are made up of sugars and phosphates, and the rungs are formed by bonded pairs of nitrogenous bases. These bases are adenine (A), guanine (G), cytosine (C), and thymine (T). An A on one chain bonds to a T on the other (thus forming an A–T ladder rung); similarly, a C on one chain bonds to a G on the other. If the bonds between the bases are broken, the two chains unwind, and free nucleotides within the cell attach themselves to the exposed bases of the now-separated chains. The free nucleotides line up along each chain according to the base-pairing rule—A bonds to T, C bonds to G. This process results in the creation of two identical DNA molecules from one original and is the method by which hereditary information is passed from one generation of cells to the next.

Gene Transcription and Translation

The sequence of bases along a strand of DNA determines the genetic code. When the product of a particular gene is needed, the portion of the DNA molecule that contains that gene will split. Through the process of transcription, a strand of RNA with bases complementary to those of the gene is created from the free nucleotides in the cell. (RNA has the base uracil [U] instead of thymine, so A and U form base pairs during RNA synthesis). This single chain of RNA, called messenger RNA (mRNA), then passes to the organelles called ribosomes, where the process of translation, or protein synthesis, takes place. During translation, a second type of RNA, transfer RNA (tRNA), matches up the nucleotides on mRNA with specific amino acids. Each set of three nucleotides codes for one amino acid. The series of amino acids built according to the sequence of nucleotides forms a polypeptide chain; all proteins are made from one or more linked polypeptide chains.

Experiments conducted in the 1940s indicated one gene being responsible for the assembly of one enzyme, or one polypeptide chain. This is known as the one gene–one enzyme hypothesis. However, since this discovery, it has been realized that not all genes encode an enzyme and that some enzymes are made up of several short polypeptides encoded by two or more genes.

Gene Regulation

Experiments have shown that many of the genes within the cells of organisms are inactive much or even all of the time. Thus, at any time, in both eukaryotes and prokaryotes, it seems that a gene can be switched on or off. The regulation of genes between eukaryotes and prokaryotes differs in important ways.

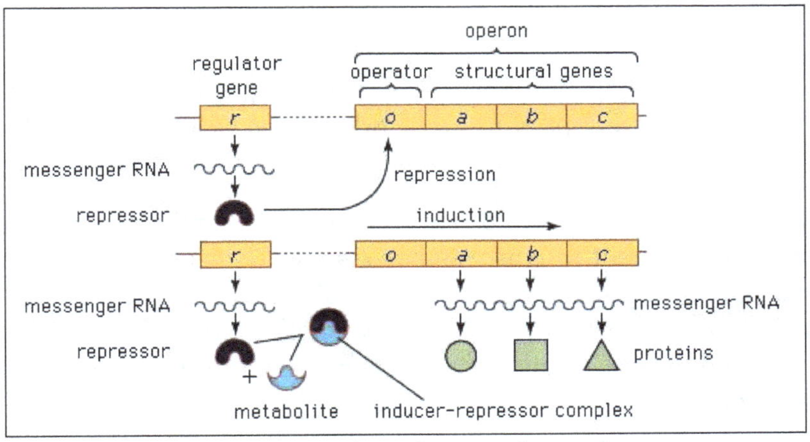

Model of the operon and its relation to the regulator gene.

The process by which genes are activated and deactivated in bacteria is well characterized. Bacteria have three types of genes: structural, operator, and regulator. Structural genes code for the synthesis of specific polypeptides. Operator genes contain the code necessary to begin the process of transcribing the DNA message of one or more structural genes into mRNA. Thus, structural genes are linked to an operator gene in a functional unit called an operon. Ultimately, the activity of the operon is controlled by a regulator gene, which produces a small protein molecule called a repressor. The repressor binds to the operator gene and prevents it from initiating the synthesis of the protein called for by the operon. The presence or absence of certain repressor molecules determines whether the operon is off or on. As mentioned, this model applies to bacteria.

The genes of eukaryotes, which do not have operons, are regulated independently. The series of events associated with gene expression in higher organisms involves multiple levels of regulation and is often influenced by the presence or absence of molecules called transcription factors. These factors influence the fundamental level of gene control, which is the rate of transcription, and may function as activators or enhancers. Specific transcription factors regulate the production of RNA from genes at certain times and in certain types of cells. Transcription factors often bind to the promoter, or regulatory region, found in the genes of higher organisms. Following transcription, introns (noncoding nucleotide sequences) are excised from the primary transcript through processes known as editing and splicing. The result of these processes is a functional strand of mRNA. For most genes this is a routine step in the production of mRNA, but in some genes there are multiple ways to splice the primary transcript, resulting in different mRNAs, which in turn result in different proteins. Some genes also are controlled at the translational and posttranslational levels.

Gene Mutations

Mutations occur when the number or order of bases in a gene is disrupted. Nucleotides can be deleted, doubled, rearranged, or replaced, each alteration having a particular effect. Mutation generally has little or no effect, but, when it does alter an organism, the change may be lethal or cause disease. A beneficial mutation will rise in frequency within a population until it becomes the norm.

GENETICS

Genetics is the study of heredity in general and of genes in particular. Genetics forms one of the central pillars of biology and overlaps with many other areas, such as agriculture, medicine, and biotechnology.

Since the dawn of civilization, humankind has recognized the influence of heredity and applied its principles to the improvement of cultivated crops and domestic animals. A Babylonian tablet more than 6,000 years old, for example, shows pedigrees of horses and indicates possible inherited characteristics. Other old carvings show cross-pollination of date palm trees. Most of the mechanisms of heredity, however, remained a mystery until the 19th century, when genetics as a systematic science began.

Genetics arose out of the identification of genes, the fundamental units responsible for heredity. Genetics may be defined as the study of genes at all levels, including the ways in which they act in the cell and the ways in which they are transmitted from parents to offspring. Modern genetics focuses on the chemical substance that genes are made of, called deoxyribonucleic acid, or DNA, and the ways in which it affects the chemical reactions that constitute the living processes within the cell. Gene action depends on interaction with the environment. Green plants, for example, have genes containing the information necessary to synthesize the photosynthetic pigment chlorophyll that gives them their green colour. Chlorophyll is synthesized in an environment containing light because the gene for chlorophyll is expressed only when it interacts with light. If a plant is placed in a dark environment, chlorophyll synthesis stops because the gene is no longer expressed.

Genetics as a scientific discipline stemmed from the work of Gregor Mendel in the middle of the 19th century. Mendel suspected that traits were inherited as discrete units, and, although he knew nothing of the physical or chemical nature of genes at the time, his units became the basis for the development of the present understanding of heredity. All present research in genetics can be traced back to Mendel's discovery of the laws governing the inheritance of traits. The word genetics was introduced in 1905 by English biologist William Bateson, who was one of the discoverers of Mendel's work and who became a champion of Mendel's principles of inheritance.

Areas of Study

Classical Genetics

Classical genetics, which remains the foundation for all other areas in genetics, is concerned primarily with the method by which genetic traits—classified as dominant (always expressed), recessive (subordinate to a dominant trait), intermediate (partially expressed), or polygenic (due to multiple genes)—are transmitted in plants and animals. These traits may be sex-linked (resulting from the action of a gene on the sex, or X, chromosome) or autosomal (resulting from the action of a gene on a chromosome other than a sex chromosome). Classical genetics began with Mendel's study of inheritance in garden peas and continues with studies of inheritance in many different plants and animals. Today a prime reason for performing classical genetics is for gene discovery—the finding and assembling of a set of genes that affects a biological property of interest.

Cytogenetics

Cytogenetics, the microscopic study of chromosomes, blends the skills of cytologists, who study the structure and activities of cells, with those of geneticists, who study genes. Cytologists discovered chromosomes and the way in which they duplicate and separate during cell division at about the same time that geneticists began to understand the behavior of genes at the cellular level. The close correlation between the two disciplines led to their combination.

Plant cytogenetics early became an important subdivision of cytogenetics because, as a general rule, plant chromosomes are larger than those of animals. Animal cytogenetics became important after the development of the so-called squash technique, in which entire cells are pressed flat on a piece of glass and observed through a microscope; the human chromosomes were numbered using this technique.

Today there are multiple ways to attach molecular labels to specific genes and chromosomes, as well as to specific RNAs and proteins, that make these molecules easily discernible from other components of cells, thereby greatly facilitating cytogenetics research.

Microbial Genetics

Microorganisms were generally ignored by the early geneticists because they are small in size and were thought to lack variable traits and the sexual reproduction necessary for a mixing of genes from different organisms. After it was discovered that microorganisms have many different physical and physiological characteristics that are amenable to study, they became objects of great interest to geneticists because of their small size and the fact that they reproduce much more rapidly than larger organisms. Bacteria became important model organisms in genetic analysis, and many discoveries of general interest in genetics arose from their study. Bacterial genetics is the centre of cloning technology.

Viral genetics is another key part of microbial genetics. The genetics of viruses that attack bacteria were the first to be elucidated. Since then, studies and findings of viral genetics have been applied to viruses pathogenic on plants and animals, including humans. Viruses are also used as vectors (agents that carry and introduce modified genetic material into an organism) in DNA technology.

Molecular Genetics

Molecular genetics is the study of the molecular structure of DNA, its cellular activities (including its replication), and its influence in determining the overall makeup of an organism. Molecular genetics relies heavily on genetic engineering (recombinant DNA technology), which can be used to modify organisms by adding foreign DNA, thereby forming transgenic organisms. Since the early 1980s, these techniques have been used extensively in basic biological research and are also fundamental to the biotechnology industry, which is devoted to the manufacture of agricultural and medical products. Transgenesis forms the basis of gene therapy, the attempt to cure genetic disease by addition of normally functioning genes from exogenous sources.

Genomics

The development of the technology to sequence the DNA of whole genomes on a routine basis has given rise to the discipline of genomics, which dominates genetics research today. Genomics is the study of the structure, function, and evolutionary comparison of whole genomes. Genomics has made it possible to study gene function at a broader level, revealing sets of genes that interact to impinge on some biological property of interest to the researcher. Bioinformatics is the computer-based discipline that deals with the analysis of such large sets of biological information, especially as it applies to genomic information.

Population Genetics

The study of genes in populations of animals, plants, and microbes provides information on past migrations, evolutionary relationships and extents of mixing among different varieties and species, and methods of adaptation to the environment. Statistical methods are used to analyze gene distributions and chromosomal variations in populations.

Population genetics is based on the mathematics of the frequencies of alleles and of genetic types in populations. For example, the Hardy-Weinberg formula, $p^2 + 2pq + q^2 = 1$, predicts the frequency of individuals with the respective homozygous dominant (AA), heterozygous (Aa), and homozygous recessive (aa) genotypes in a randomly mating population. Selection, mutation, and random changes can be incorporated into such mathematical models to explain and predict the course of evolutionary change at the population level. These methods can be used on alleles of known phenotypic effect, such as the recessive allele for albinism, or on DNA segments of any type of known or unknown function.

Human population geneticists have traced the origins and migration and invasion routes of modern humans, Homo sapiens. DNA comparisons between the present peoples on the planet have pointed to an African origin of Homo sapiens. Tracing specific forms of genes has allowed geneticists to deduce probable migration routes out of Africa to the areas colonized today. Similar studies show to what degree present populations have been mixed by recent patterns of travel.

Behavior Genetics

Another aspect of genetics is the study of the influence of heredity on behavior. Many aspects of animal behavior are genetically determined and can therefore be treated as similar to other biological properties. This is the subject material of behavior genetics, whose goal is to determine which genes control various aspects of behavior in animals. Human behavior is difficult to analyze because of the powerful effects of environmental factors, such as culture. Few cases of genetic determination of complex human behavior are known. Genomics studies provide a useful way to explore the genetic factors involved in complex human traits such as behavior.

Human Genetics

Some geneticists specialize in the hereditary processes of human genetics. Most of the emphasis is on understanding and treating genetic disease and genetically influenced ill health, areas collectively known as medical genetics. One broad area of activity is laboratory research dealing with the mechanisms of human gene function and malfunction and investigating pharmaceutical and other types of treatments. Since there is a high degree of evolutionary conservation between organisms, research on model organisms—such as bacteria, fungi, and fruit flies (Drosophila)—which are easier to study, often provides important insights into human gene function.

Many single-gene diseases, caused by mutant alleles of a single gene, have been discovered. Two well-characterized single-gene diseases include phenylketonuria (PKU) and Tay-Sachs disease. Other diseases, such as heart disease, schizophrenia, and depression, are thought to have more complex heredity components that involve a number of different genes. These diseases are the focus of a great deal of research that is being carried out today.

Another broad area of activity is clinical genetics, which centres on advising parents of the likelihood of their children being affected by genetic disease caused by mutant genes and abnormal chromosome structure and number. Such genetic counseling is based on examining individual and family medical records and on diagnostic procedures that can detect unexpressed, abnormal forms of genes. Counseling is carried out by physicians with a particular interest in this area or by specially trained nonphysicians.

Methods in Genetics

Experimental Breeding

Genetically diverse lines of organisms can be crossed in such a way to produce different combinations of alleles in one line. For example, parental lines are crossed, producing an F1 generation, which is then allowed to undergo random mating to produce offspring that have purebreeding genotypes (i.e., AA, bb, cc, or DD). This type of experimental breeding is the origin of new plant and animal lines, which are an important part of making laboratory stocks for basic research. When applied to commerce, transgenic commercial lines produced experimentally are called genetically modified organisms (GMOs). Many of the plants and animals used by humans today (e.g., cows, pigs, chickens, sheep, wheat, corn (maize), potatoes, and rice) have been bred in this way.

Cytogenetic Techniques

Cytogenetics focuses on the microscopic examination of genetic components of the cell, including chromosomes, genes, and gene products. Older cytogenetic techniques involve placing cells in paraffin wax, slicing thin sections, and preparing them for microscopic study. The newer and faster squash technique involves squashing entire cells and studying their contents. Dyes that selectively stain various parts of the cell are used; the genes, for example, may be located by selectively staining the DNA of which they are composed. Radioactive and fluorescent tags are valuable in determining the location of various genes and gene products in the cell. Tissue-culture techniques may be used to grow cells before squashing; white blood cells can be grown from samples of human blood and studied with the squash technique. One major application of cytogenetics in humans is in diagnosing abnormal chromosomal complements such as Down syndrome (caused by an extra copy of chromosome 21) and Klinefelter syndrome (occurring in males with an extra X chromosome). Some diagnosis is prenatal, performed on cell samples from amniotic fluid or the placenta.

Biochemical Techniques

Biochemistry is carried out at the cellular or subcellular level, generally on cell extracts. Biochemical methods are applied to the main chemical compounds of genetics—notably DNA, RNA, and protein. Biochemical techniques are used to determine the activities of genes within cells and to analyze substrates and products of gene-controlled reactions. In one approach, cells are ground up and the substituent chemicals are fractionated for further analysis. Special techniques (e.g., chromatography and electrophoresis) are used to separate the components of proteins so that inherited differences in their structures can be revealed. For example, more than 100 different kinds of human hemoglobin molecules have been identified. Radioactively tagged compounds are valuable in studying the biochemistry of whole cells. For example, thymine is a compound found only in DNA; if radioactive thymine is placed in a tissue-culture medium in which cells are growing, genes use it to duplicate themselves. When cells containing radioactive thymine are analyzed, the results show that, during duplication, the DNA molecule splits in half, and each half synthesizes its missing components.

Chemical tests are used to distinguish certain inherited conditions of humans; e.g., urinalysis and blood analysis reveal the presence of certain inherited abnormalities—phenylketonuria (PKU), cystinuria, alkaptonuria, gout, and galactosemia. Genomics has provided a battery of diagnostic tests that can be carried out on an individual's DNA. Some of these tests can be applied to fetuses in utero.

Physiological Techniques

Physiological techniques, directed at exploring functional properties or organisms, are also used in genetic investigations. In microorganisms, most genetic variations involve some important cell function. Some strains of one bacterium (Escherichia coli), for example, are able to synthesize the vitamin thiamin from simple compounds; others, which lack an enzyme necessary for this synthesis, cannot survive unless thiamin is already present. The two strains can be distinguished by placing them on a thiamin-free mixture: those that grow have the gene for the enzyme, those that fail to grow do not. The technique also is applied to human cells, since many inherited human abnormalities are caused by a faulty gene that fails to produce a vital enzyme; albinism, which results from an inability to produce the pigment melanin in the skin, hair, or iris of the eyes, is an example of an enzyme deficiency in man.

Molecular Techniques

Although overlapping with biochemical techniques, molecular genetics techniques are deeply involved with the direct study of DNA. This field has been revolutionized by the invention of recombinant DNA technology. The DNA of any gene of interest from a donor organism (such as a human) can be cut out of a chromosome and inserted into a vector to make recombinant DNA, which can then be amplified and manipulated, studied, or used to modify the genomes of other organisms by transgenesis. A fundamental step in recombinant DNA technology is amplification. This is carried out by inserting the recombinant DNA molecule into a bacterial cell, which replicates and produces many copies of the bacterial genome and the recombinant DNA molecule (constituting a DNA clone). A collection of large numbers of clones of recombinant donor DNA molecules is called a genomic library. Such libraries are the starting point for sequencing entire genomes such as the human genome. Today genomes can be scanned for small molecular variants called single nucleotide polymorphisms, or SNPs ("snips"), which act as chromosomal tags to associated specific regions of DNA that have a property of interest and may be involved in a human disease or disorder.

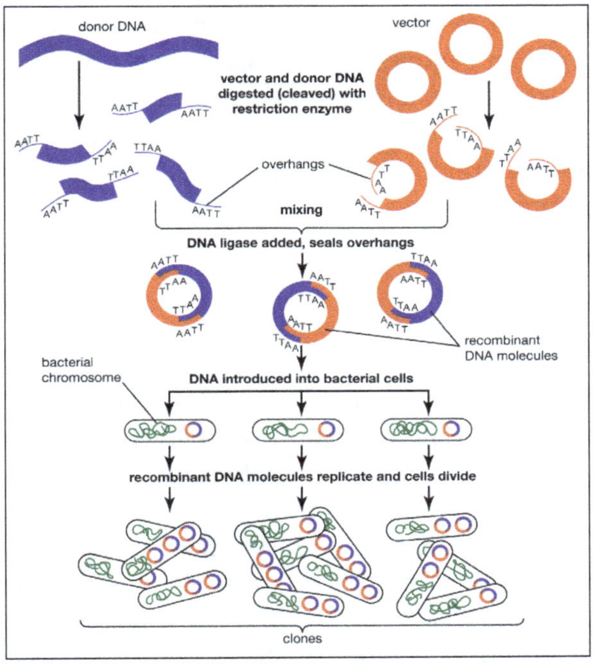

Steps involved in the engineering of a recombinant DNA molecule.

Immunological Techniques

Many substances (e.g., proteins) are antigenic; i.e., when introduced into a vertebrate body, they stimulate the production of specific proteins called antibodies. Various antigens exist in red blood cells, including those that make up the major blood groups of man (A, B, AB, O). These and other antigens are genetically determined; their study constitutes immunogenetics. Blood antigens of man include inherited variations, and the particular combination of antigens in an individual is almost as unique as fingerprints and has been used in such areas as paternity testing (although this approach has been largely supplanted by DNA-based techniques).

Immunological techniques are used in blood group determinations in blood transfusions, in organ transplants, and in determining Rhesus incompatibility in childbirth. Specific antigens of the human leukocyte antigen (HLA) genes are correlated with human diseases and disease predispositions. Antibodies also have a genetic basis, and their seemingly endless ability to match any antigen presented is based on special types of DNA shuffling processes between antibody genes. Immunology is also useful in identifying specific recombinant DNA clones that synthesize a specific protein of interest.

Mathematical Techniques

Because much of genetics is based on quantitative data, mathematical techniques are used extensively in genetics. The laws of probability are applicable to crossbreeding and are used to predict frequencies of specific genetic constitutions in offspring. Geneticists also use statistical methods to determine the significance of deviations from expected results in experimental analyses. In addition, population genetics is based largely on mathematical logic—for example, the Hardy-Weinberg equilibrium and its derivatives.

Bioinformatics uses computer-centred statistical techniques to handle and analyze the vast amounts of information accumulating from genome sequencing projects. The computer program scans the DNA looking for genes, determining their probable function based on other similar genes, and comparing different DNA molecules for evolutionary analysis. Bioinformatics has made possible the discipline of systems biology, treating and analyzing the genes and gene products of cells as a complete and integrated system.

Applied Genetics

Medicine

Genetic techniques are used in medicine to diagnose and treat inherited human disorders. Knowledge of a family history of conditions such as cancer or various disorders may indicate a hereditary tendency to develop these afflictions. Cells from embryonic tissues reveal certain genetic abnormalities, including enzyme deficiencies, that may be present in newborn babies, thus permitting early treatment. Many countries require a blood test of newborn babies to determine the presence of an enzyme necessary to convert an amino acid, phenylalanine, into simpler products. Phenylketonuria (PKU), which results from lack of the enzyme, causes permanent brain damage if not treated soon after birth. Many different types of human genetic diseases can be detected in embryos as young as 12 weeks; the procedure involves removal and testing of a small amount of fluid from around the embryo (called amniocentesis) or of tissue from the placenta (called chorionic villus sampling).

Gene therapy is based on modification of defective genotypes by adding functional genes made through recombinant DNA technology. Bioinformatics is being used to "mine" the human genome for gene products that might be candidates for designer pharmaceutical drugs.

Agriculture and Animal Husbandry

Agriculture and animal husbandry apply genetic techniques to improve plants and animals. Breeding analysis and transgenic modification using recombinant DNA techniques are routinely used. Animal breeders use artificial insemination to propagate the genes of prize bulls. Prize cows can transmit their genes to hundreds of offspring by hormone treatment, which stimulates the release of many eggs that are collected, fertilized, and transplanted to foster mothers. Several types of mammals can be cloned, meaning that multiple identical copies can be produced of certain desirable types.

Plant geneticists use special techniques to produce new species, such as hybrid grains (i.e., produced by crossing wheat and rye), and plants resistant to destruction by insect and fungal pests.

Plant breeders use the techniques of budding and grafting to maintain desirable gene combinations originally obtained from crossbreeding. Transgenic plant cells can be made into plants by growing the cells on special hormones. The use of the chemical compound colchicine, which causes chromosomes to double in number, has resulted in many new varieties of fruits, vegetables, and flowers. Many transgenic lines of crop plants are commercially advantageous and are being introduced into the market.

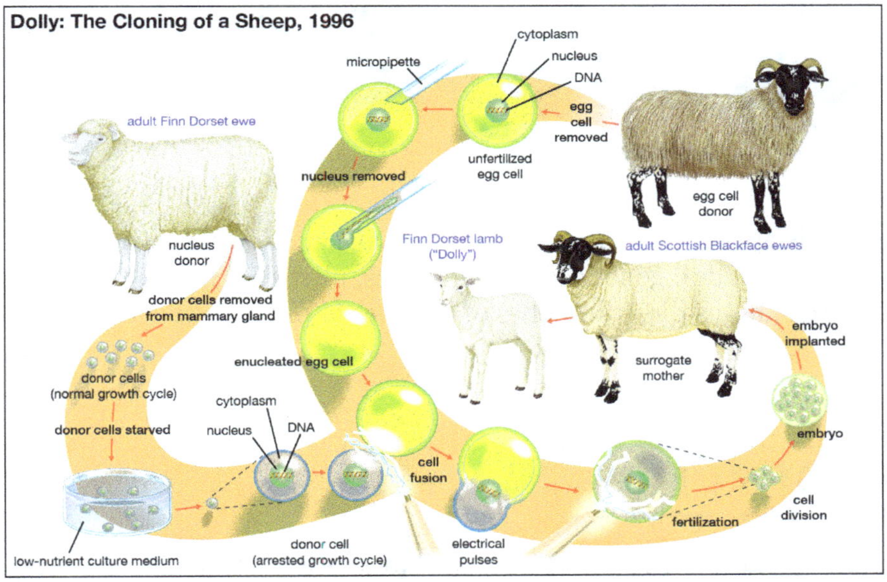

Dolly the sheep was successfully cloned in 1996 by fusing the nucleus from a mammary-gland cell of a Finn Dorset ewe into an enucleated egg cell taken from a Scottish Blackface ewe. Carried to term in the womb of another Scottish Blackface ewe, Dolly was a genetic copy of the Finn Dorset ewe.

Industry

Various industries employ geneticists; the brewing industry, for example, may use geneticists to improve the strains of yeast that produce alcohol. The pharmaceutical industry has developed strains of molds, bacteria, and other microorganisms high in antibiotic yield. Penicillin and cyclosporin from fungi, and streptomycin and ampicillin from bacteria, are some examples.

Biotechnology, based on recombinant DNA technology, is now extensively used in industry. "Designer" lines of transgenic bacteria, animals, or plants capable of manufacturing some commercial product are made and used routinely. Such products include pharmaceutical drugs and industrial chemicals such as citric acid.

Branches of Genetics

CHAPTER 2

There are numerous branches of genetics such as human genetics, behavior genetics, cytogenetics, molecular genetics, population genetics, plant genetics, microbial genetics and classical genetics. The chapter closely examines these branches of genetics to provide an extensive understanding of the subject.

HUMAN GENETICS

Human genetics is the study of the inheritance of characteristics by children from parents. Inheritance in humans does not differ in any fundamental way from that in other organisms.

The study of human heredity occupies a central position in genetics. Much of this interest stems from a basic desire to know who humans are and why they are as they are. At a more practical level, an understanding of human heredity is of critical importance in the prediction, diagnosis, and treatment of diseases that have a genetic component. The quest to determine the genetic basis of human health has given rise to the field of medical genetics. In general, medicine has given focus and purpose to human genetics, so the terms medical genetics and human genetics are often considered synonymous.

Human Chromosomes

A new era in cytogenetics, the field of investigation concerned with studies of the chromosomes, began in 1956 with the discovery by Jo Hin Tjio and Albert Levan that human somatic cells contain 23 pairs of chromosomes. Since that time the field has advanced with amazing rapidity and has demonstrated that human chromosome aberrations rank as major causes of fetal death and of tragic human diseases, many of which are accompanied by mental retardation. Since the chromosomes can be delineated only during mitosis, it is necessary to examine material in which there are many dividing cells. This can usually be accomplished by culturing cells from the blood or skin, since only the bone marrow cells (not readily sampled except during serious bone marrow disease such as leukemia) have sufficient mitoses in the absence of artificial culture. After growth, the cells are fixed on slides and then stained with a variety of DNA-specific stains that permit the delineation and identification of the chromosomes. The Denver system of chromosome classification, established in 1959, identified the chromosomes by their length and the position of the centromeres. Since then the method has been improved by the use of special staining techniques that impart unique light and dark bands to each chromosome. These bands permit the identification of chromosomal regions that are duplicated, missing, or transposed to other chromosomes.

Micrographs showing the karyotypes (i.e., the physical appearance of the chromosome) of a male and a female have been produced. In a typical micrograph the 46 human chromosomes (the diploid number) are arranged in homologous pairs, each consisting of one maternally derived and one paternally derived member. The chromosomes are all numbered except for the X and the Y chromosomes, which are the sex chromosomes. In humans, as in all mammals, the normal female has two X chromosomes and the normal male has one X chromosome and one Y chromosome. The female is thus the homogametic sex, as all her gametes normally have one X chromosome. The male is heterogametic, as he produces two types of gametes—one type containing an X chromosome and the other containing a Y chromosome. There is good evidence that the Y chromosome in humans, unlike that in Drosophila, is necessary (but not sufficient) for maleness.

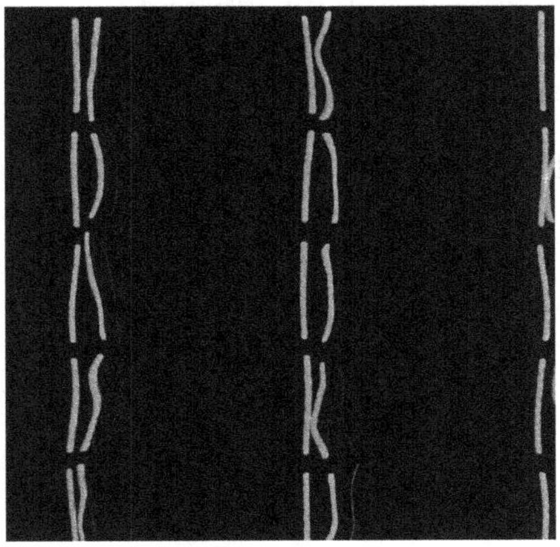

Strands of human chromosomes.

Fertilization, Sex Determination and Differentiation

A human individual arises through the union of two cells, an egg from the mother and a sperm from the father. Human egg cells are barely visible to the naked eye. They are shed, usually one at a time, from the ovary into the oviducts (fallopian tubes), through which they pass into the uterus. Fertilization, the penetration of an egg by a sperm, occurs in the oviducts. This is the main event of sexual reproduction and determines the genetic constitution of the new individual.

Human sex determination is a genetic process that depends basically on the presence of the Y chromosome in the fertilized egg. This chromosome stimulates a change in the undifferentiated gonad into that of the male (a testicle). The gonadal action of the Y chromosome is mediated by a gene located near the centromere; this gene codes for the production of a cell surface molecule called the H-Y antigen. Further development of the anatomic structures, both internal and external, that are associated with maleness is controlled by hormones produced by the testicle. The sex of an individual can be thought of in three different contexts: chromosomal sex, gonadal sex, and anatomic sex. Discrepancies between these, especially the latter two, result in the development of individuals with ambiguous sex, often called hermaphrodites. The phenomenon of homosexuality is of uncertain cause and is unrelated to the above sex-determining factors. It is of interest that in the absence of a male gonad (testicle) the internal and external sex anatomy is always female, even in the absence of a female ovary. A female without ovaries will, of course, be infertile and will not experience any of the female developmental changes normally associated with puberty. Such a female will often have Turner syndrome.

If X-containing and Y-containing sperm are produced in equal numbers, then according to simple chance one would expect the sex ratio at conception (fertilization) to be half boys and half girls, or 1 : 1. Direct observation of sex ratios among newly fertilized human eggs is not yet feasible, and sex-ratio data are usually collected at the time of birth. In almost all human populations of newborns, there is a slight excess of males; about 106 boys are born for every100 girls. Throughout life, however, there is a slightly greater mortality of males; this slowly alters the sex ratio until, beyond the age of about 50 years, there is an excess of females. Studies indicate that male embryos suffer a relatively greater degree of prenatal mortality, so the sex ratio at conception might be expected to favour males even more than the 106 : 100 ratio observed at birth would suggest. Firm explanations for the apparent excess of male conceptions have not been established; it is possible that Y-containing sperm survive better within the female reproductive tract, or they may be a little more successful in reaching the egg in order to fertilize it. In any case, the sex differences are small, the statistical expectation for a boy (or girl) at any single birth still being close to one out of two.

During gestation—the period of nine months between fertilization and the birth of the infant—a remarkable series of developmental changes occur. Through the process of mitosis, the total number of cells changes from 1 (the fertilized egg) to about 2×10^{11}. In addition, these cells differentiate into hundreds of different types with specific functions (liver cells, nerve cells, muscle cells, etc.). A multitude of regulatory processes, both genetically and environmentally controlled, accomplish this differentiation. Elucidation of the exquisite timing of these processes remains one of the great challenges of human biology.

Immunogenetics

Immunity is the ability of an individual to recognize the "self" molecules that make up one's own body and to distinguish them from such "nonself" molecules as those found in infectious microorganisms and toxins. This process has a prominent genetic component. Knowledge of the genetic and molecular basis of the mammalian immune system has increased in parallel with the explosive advances made in somatic cell and molecular genetics.

There are two major components of the immune system, both originating from the same precursor "stem" cells. The bursa component provides B lymphocytes, a class of white blood cells that, when appropriately stimulated, differentiate into plasma cells. These latter cells produce circulating soluble proteins called antibodies or immunoglobulins. Antibodies are produced in response to substances called antigens, most of which are foreign proteins or polysaccharides. An antibody molecule can recognize a specific antigen, combine with it, and initiate its destruction. This so-called humoral immunity is accomplished through a complicated series of interactions with other molecules and cells; some of these interactions are mediated by another group of lymphocytes, the T lymphocytes, which are derived from the thymus gland. Once a B lymphocyte has been exposed to a specific antigen, it "remembers" the contact so that future exposure will cause an accelerated and magnified immune reaction. This is a manifestation of what has been called immunological memory.

The thymus component of the immune system centres on the thymus-derived T lymphocytes. In addition to regulating the B cells in producing humoral immunity, the T cells also directly attack cells that display foreign antigens. This process, called cellular immunity, is of great importance in protecting the body against a variety of viruses as well as cancer cells. Cellular immunity is also the chief cause of the rejection of organ transplants. The T lymphocytes provide a complex network consisting of a series of helper cells (which are antigen-specific), amplifier cells, suppressor cells, and cytotoxic (killer) cells, all of which are important in immune regulation.

The Genetics of Antibody Formation

One of the central problems in understanding the genetics of the immune system has been in explaining the genetic regulation of antibody production. Immunobiologists have demonstrated that the system can produce well over one million specific antibodies, each corresponding to a particular antigen. It would be difficult to envisage that each antibody is encoded by a separate gene; such an arrangement would require a disproportionate share of the entire human genome. Recombinant DNA analysis has illuminated the mechanisms by which a limited number of immunoglobulin genes can encode this vast number of antibodies.

Each antibody molecule consists of several different polypeptide chains—the light chains (L) and the longer heavy chains (H). The latter determine to which of five different classes (IgM, IgG, IgA, IgD, or IgE) an immunoglobulin belongs. Both the L and H chains are unique among proteins in that they contain constant and variable parts. The constant parts have relatively identical amino acid sequences in any given antibody. The variable parts, on the other hand, have different amino acid sequences in each antibody molecule. It is the variable parts, then, that determine the specificity of the antibody.

Recombinant DNA studies of immunoglobulin genes in mice have revealed that the light-chain genes are encoded in four separate parts in germ-line DNA: a leader segment (L), a variable segment (V), a joining segment (J), and a constant segment (C). These segments are widely separated in the DNA of an embryonic cell, but in a mature B lymphocyte they are found in relative proximity (albeit separated by introns). The mouse has more than 200 light-chain variable region genes, only one of which will be incorporated into the proximal sequence that codes for the antibody production in a given B lymphocyte. Antibody diversity is greatly enhanced by this system, as the V and J segments rearrange and assort randomly in each B-lymphocyte precursor cell. The mechanisms by which this DNA rearrangement takes place are not clear, but transposons are undoubtedly involved. Similar combinatorial processes take place in the genes that code for the heavy chains; furthermore, both the light-chain and heavy-chain genes can undergo somatic mutations to create new antibody-coding sequences. The net effect of these combinatorial and mutational processes enables the coding of millions of specific antibody molecules from a limited number of genes. It should be stressed, however, that each B lymphocyte can produce only one antibody. It is the B lymphocyte population as a whole that produces the tremendous variety of antibodies in humans and other mammals.

Plasma cell tumours (myelomas) have made it possible to study individual antibodies, since these tumours, which are descendants of a single plasma cell, produce one antibody in abundance. Another method of obtaining large amounts of a specific antibody is by fusing a B lymphocyte with a rapidly growing cancer cell. The resultant hybrid cell, known as a hybridoma, multiplies rapidly in culture. Since the antibodies obtained from hybridomas are produced by clones derived from a single lymphocyte, they are called monoclonal antibodies.

The Genetics of Cellular Immunity

As has been stated, cellular immunity is mediated by T lymphocytes that can recognize infected body cells, cancer cells, and the cells of a foreign transplant. The control of cellular immune reactions is provided by a linked group of genes, known as the major histocompatibility complex (MHC). These genes code for the major histocompatibility antigens, which are found on the surface of almost all nucleated somatic cells. The major histocompatibility antigens were first discovered on the leukocytes (white blood cells) and are therefore usually referred to as the HLA (human leukocyte group A) antigens.

The advent of the transplantation of human organs in the 1950s made the question of tissue compatibility between donor and recipient of vital importance, and it was in this context that the HLA antigens and the MHC were elucidated. Investigators found that the MHC resides on the short arm of chromosome 6, on four closely associated sites designated HLA-A, HLA-B, HLA-C, and HLA-D. Each locus is highly polymorphic; i.e., each is represented by a great many alleles within the human gene pool. These alleles, like those of the ABO blood group system, are expressed in co-dominant fashion. Because of the large number of alleles at each HLA locus, there is an extremely low probability of any two individuals (other than siblings) having identical HLA genotypes. (Since a person inherits one chromosome 6 from each parent, siblings have a 25 percent probability of having received the same paternal and maternal chromosomes 6 and thus of being HLA matched).

Although HLA antigens are largely responsible for the rejection of organ transplants, it is obvious that the MHC did not evolve to prevent the transfer of organs from one person to another. Indeed, information obtained from the histocompatibility complex in the mouse (which is very similar in its genetic organization to that of the human) suggests that a primary function of the HLA antigens is to regulate the number of specific cytotoxic T killer cells, which have the ability to destroy virus-infected cells and cancer cells.

Genetics of Human Blood

More is known about the genetics of the blood than about any other human tissue. One reason for this is that blood samples can be easily secured and subjected to biochemical analysis without harm or major discomfort to the person being tested. Perhaps a more cogent reason is that many chemical properties of human blood display relatively simple patterns of inheritance.

Blood Types

Certain chemical substances within the red blood cells may serve as antigens. When cells that contain specific antigens are introduced into the body of an experimental animal such as a rabbit, the animal responds by producing antibodies in its own blood.

In addition to the ABO and MN systems, geneticists have identified about 14 blood-type gene systems associated with other chromosomal locations. The best known of these is the Rh system. The Rh antigens are of particular importance in human medicine. Curiously, however, their existence was discovered in monkeys. When blood from the rhesus monkey (hence the designation Rh) is injected into rabbits, the rabbits produce so-called Rh antibodies that will agglutinate not only the red blood cells of the monkey but the cells of a large proportion of human beings as well. Some people (Rh-negative individuals), however, lack the Rh antigen; the proportion of such persons varies from one human population to another. Akin to data concerning the ABO system, the evidence for Rh genes indicates that only a single chromosome locus (called r) is involved and is located on chromosome 1. At least 35 Rh alleles are known for the r location; basically the Rh-negative condition is recessive.

A medical problem may arise when a woman who is Rh-negative carries a fetus that is Rh-positive. The first such child may have no difficulty, but later similar pregnancies may produce severely anemic newborn infants. Exposure to the red blood cells of the first Rh-positive fetus appears to immunize the Rh-negative mother, that is, she develops antibodies that may produce permanent (sometimes fatal) brain damage in any subsequent Rh-positive fetus. Damage arises from the scarcity of oxygen reaching the fetal brain because of the severe destruction of red blood cells. Measures are available for avoiding the severe effects of Rh incompatibility by transfusions to the fetus within the uterus; however, genetic counselling before conception is helpful so that the mother can receive Rh immunoglobulin immediately after her first and any subsequent pregnancies involving an Rh-positive fetus. This immunoglobulin effectively destroys the fetal red blood cells before the mother's immune system is stimulated. The mother thus avoids becoming actively immunized against the Rh antigen and will not produce antibodies that could attack the red blood cells of a future Rh-positive fetus.

Serum Proteins

Human serum, the fluid portion of the blood that remains after clotting, contains various proteins that have been shown to be under genetic control. Study of genetic influences has flourished since the development of precise methods for separating and identifying serum proteins. These move at different rates under the impetus of an electrical field (electrophoresis), as do proteins from many other sources (e.g., muscle or nerve). Since the composition of a protein is specified by the structure of its corresponding gene, biochemical studies based on electrophoresis permit direct study of tissue substances that are only a metabolic step or two away from the genes themselves.

Electrophoretic studies have revealed that at least one-third of the human serum proteins occur in variant forms. Many of the serum proteins are polymorphic, occurring as two or more variants with a frequency of not less than 1 percent each in a population. Patterns of polymorphic serum protein variants have been used to determine whether twins are identical (as in assessing compatibility for organ transplants) or whether two individuals are related (as in resolving paternity suits). Whether the different forms have a selective advantage is not generally known.

Much attention in the genetics of substances in the blood has been centred on serum proteins called haptoglobins, transferrins (which transport iron), and gamma globulins (a number of which are known to immunize against infectious diseases). Haptoglobins appear to relate to two common alleles at a single chromosome locus; the mode of inheritance of the other two seems more complicated, about 18 kinds of transferrins having been described. Like blood-cell antigen genes, serum-protein genes are distributed worldwide in the human population in a way that permits their use in tracing the origin and migration of different groups of people.

Hemoglobin

Hundreds of variants of hemoglobin have been identified by electrophoresis, but relatively few are frequent enough to be called polymorphisms. Of the polymorphisms, the alleles for sickle-cell and thalassemia hemoglobins produce serious disease in homozygotes, whereas others (hemoglobins C, D, and E) do not. The sickle-cell polymorphism confers a selective advantage on the heterozygote living in a malarial environment; the thalassemia polymorphism provides a similar advantage.

Influence of the Environment

Gene expression occurs only after modification by the environment. A good example is the recessively inherited disease called galactosemia, in which the enzyme necessary for the metabolism of galactose—a component of milk sugar—is defective. The sole source of galactose in the infant's diet is milk, which in this instance is toxic. The treatment of this most serious disease in the neonate is to remove all natural forms of milk from the diet (environmental manipulation) and to substitute a synthetic milk lacking galactose. The infant will then develop normally but will never be able to tolerate foods containing lactose. If milk was not a major part of the infant's diet, however, the mutant gene would never be able to express itself, and galactosemia would be unknown.

Another way of saying this is that no trait can exist or become actual without an environmental contribution. Thus, the old question of which is more important, heredity or environment, is without meaning. Both nature (heredity) and nurture (environment) are always important for every human attribute.

But this is not to say that the separate contributions of heredity and environment are equivalent for each characteristic. Dark pigmentation of the iris of the eye, for example, is under hereditary control in that one or more genes specify the synthesis and deposition in the iris of the pigment (melanin). This is one characteristic that is relatively independent of such environmental factors as diet or climate; thus, individual differences in eye colour tend to be largely attributable to hereditary factors rather than to ordinary environmental change.

On the other hand, it is unwarranted to assume that other traits (such as height, weight, or intelligence) are as little affected by environment as is eye colour. It is very easy to gather information that tall parents tend, on the average, to have tall children (and that short parents tend to produce short children), properly indicating a hereditary contribution to height. Nevertheless, it is equally manifest that growth can be stunted in the environmental absence of adequate nutrition. The dilemma arises that only the combined, final result of this nature-nurture interaction can be directly observed. There is no accurate way (in the case of a single individual) to gauge the separate contributions of heredity and environment to such a characteristic as height. An inferential way out of this dilemma is provided by studies of twins.

Fraternal Twins

Usually a fertile human female produces a single egg about once a month. Should fertilization occur (a zygote is formed), growth of the individual child normally proceeds after the fertilized egg has become implanted in the wall of the uterus (womb). In the unusual circumstance that two unfertilized eggs are simultaneously released by the ovaries, each egg may be fertilized by a different sperm cell at about the same time, become implanted, and grow, to result in the birth of twins.

Twins formed from separate eggs and different sperm cells can be of the same or of either sex. No matter what their sex, they are designated as fraternal twins. This terminology is used to emphasize that fraternal twins are genetically no more alike than are siblings (brothers or sisters) born years apart. Basically they differ from ordinary siblings only in having grown side by side in the womb and in having been born at approximately the same time.

Identical Twins

In a major nonfraternal type of twinning, only one egg is fertilized, but during the cleavage of this single zygote into two cells, the resulting pair somehow become separated. Each of the two cells may implant in the uterus separately and grow into a complete, whole individual. In laboratory studies with the zygotes of many animal species, it has been found that in the two-cell stage (and later) a portion of the embryo, if separated under the microscope by the experimenter, may develop into a perfect, whole individual. Such splitting occurs spontaneously at the four-cell stage in some organisms (e.g., the armadillo) and has been accomplished experimentally with the embryos of salamanders, among others.

The net result of splitting at an early embryonic stage may be to produce so-called identical twins. Since such twins derive from the same fertilized egg, the hereditary material from which they originate is absolutely identical in every way, down to the last gene locus. While developmental and genetic differences between one "identical" twin and another still may arise through a number of processes (e.g., mutation), these twins are always found to be of the same sex. They are often breathtakingly similar in appearance, frequently down to very fine anatomic and biochemical details (although their fingerprints are differentiable).

Diagnosis of Twin Types

Since the initial event in the mother's body (either splitting of a single egg or two separate fertilizations) is not observed directly, inferential means are employed for diagnosing a set of twins as fraternal or identical. The birth of fraternal twins is frequently characterized by the passage of two separate afterbirths. In many instances, identical twins are followed by only a single afterbirth, but exceptions to this phenomenon are so common that this is not a reliable method of diagnosis.

The most trustworthy method for inferring twin type is based on the determination of genetic similarity. By selecting those traits that display the least variation attributable to environmental influences (such as eye colour and blood types), it is feasible, if enough separate chromosome loci are considered, to make the diagnosis of twin type with high confidence. HLA antigens, which, as stated above, are very polymorphic, have become most useful in this regard.

Inferences from Twin Studies

Metric (Quantitative) Traits

By measuring the heights of a large number of ordinary siblings (brothers and sisters) and of twin pairs, it may be shown that the average difference between identical twins is less than half the difference for all other siblings. Any average differences between groups of identical twins are attributable with considerable confidence to the environment. Thus, since the sample of identical twins who were reared apart (in different homes) differed little in height from identicals who were raised together, it appears that environmental-genetic influences on that trait tended to be similar for both groups.

Yet, the data for like-sexed fraternal twins reveal a much greater average difference in height (about the same as that found between ordinary siblings reared in the same home at different ages). Apparently the fraternal twins were more dissimilar than identicals (even though reared together) because the fraternals differed more from each other in genotype. This emphasizes the great genetic similarity between identicals. Such studies can be particularly enlightening when the effects of individual genes are obscured or distorted by the influence of environmental factors on quantitative (measurable) traits (e.g., height, weight, and intelligence).

Any trait that can be objectively measured in identical and fraternal twins can be scrutinized for the particular combination of hereditary and environmental influences that impinge upon it. The effect of environment on identical twins reared apart is suggested by their relatively great average difference in body weight as compared with identical twins reared together. Weight appears to be more strongly modified by environmental variables than is height.

Study of comparable characteristics among farm animals and plants suggests that such quantitative human traits as height and weight are affected by allelic differences at a number of chromosome locations—that they are not simply affected by genes at a single locus. Investigation of these gene systems with multiple locations (polygenic systems) is carried out largely through selective-breeding experiments among large groups of plants and lower animals. Human beings select their mates in a much freer fashion, of course, and polygenic studies among people are thus severely limited.

Intelligence is a very complex human trait, the genetics of which has been a subject of controversy for some time. Much of the controversy arises from the fact that intelligence is so difficult to define. Information has been based almost entirely on scores on standardized IQ tests constructed by psychologists; in general, such tests do not take into account cultural, environmental, and educational differences. As a result, the working definition of intelligence has been "the general factor common to a large number of diverse cognitive (IQ) tests." Even roughly measured as IQ, intelligence shows a strong contribution from the environment. Fraternal twins, however, show relatively great dissimilarity in IQ, suggesting an important contribution from heredity as well. In fact, it has been estimated that, on the average, between 60 and 80 percent of the variance in IQ test scores could be genetic. It is important to note that intelligence is polygenically inherited and that it has the highest degree of assortative mating of any trait; in other words, people tend to mate with people having similar IQs. Moreover, twin studies involving psychological traits should be viewed with caution; for example, since identical twins tend to be singled out for special attention, their environment should not be considered equivalent even to that of other children raised in their own family.

Since the time of Galton, generalizations have been repeatedly made about racial differences in intelligence, with claims of genetic superiority of some races over others. These generalizations fail to recognize that races are composed of individuals, each of whom has a unique genotype made up by genes shared with other humans, and that the sources of intraracial variation are more numerous than those producing interracial differences.

Other Traits

For traits of a more qualitative (all-or-none) nature, the twin method can also be used in efforts to assess the degree of hereditary contribution. Such investigations are based on an examination of cases in which at least one member of the twin pair shows the trait. It was found in one study, for example, that in about 80 percent of all identical twin pairs in which one twin shows symptoms of the psychiatric disorder called schizophrenia, the other member of the pair also shows the symptoms; that is, the two are concordant for the schizophrenic trait. In the remaining 20 percent, the twins are discordant; that is, one lacks the trait. Since identical twins often have similar environments, this information by itself does not distinguish between the effects of heredity and environment. When pairs of like-sexed fraternal twins reared together are studied, however, the degree of concordance for schizophrenia is very much lower—only about 15 percent.

Schizophrenia thus clearly develops much more easily in some genotypes than in others; this indicates a strong hereditary predisposition to the development of the trait. Schizophrenia also serves as a good example of the influence of environmental factors, since concordance for the condition does not appear in 100 percent of identical twins.

Studies of concordance and discordance between identical and fraternal twins have been carried out for many other human characteristics. It has, for example, been known for many years that tuberculosis is a bacterial infection of environmental origin. Yet identical twins raised in the same home show concordance for the disease far more often than do fraternal twins. This finding seems to be explained by the high degree of genetic similarity between the identical twins. While the tuberculosis germ is not inherited, heredity does seem to make one more (or less) susceptible to this particular infection. Thus, the genes of one individual may provide the chemical basis for susceptibility to a disease, while the genes of another may fail to do so.

Indeed, there seem to be genetic differences between disease germs themselves that result in differences in their virulence. Thus, whether a genetically susceptible person actually develops a disease also depends in part on the heredity of the particular strain of bacteria or virus with which he or she must cope. Consequently, unless environmental factors such as these are adequately evaluated, the conclusions drawn from susceptibility studies can be unfortunately misleading.

The expression of the genotype can always be modified by the environment. It can be argued that all human illnesses have a genetic component and that the basis of all medical therapy is environmental modification. Specifically, this is the hope for the management of genetic diseases. The more that can be learned about the basic molecular and cellular dysfunctions associated with such diseases, the more amenable they will be to environmental manipulation.

Human Evolutionary Genetics

Human evolutionary genetics studies how one human genome differs from another human genome, the evolutionary past that gave rise to the human genome, and its current effects. Differences between genomes have anthropological, medical, historical and forensic implications and applications. Genetic data can provide important insights into human evolution.

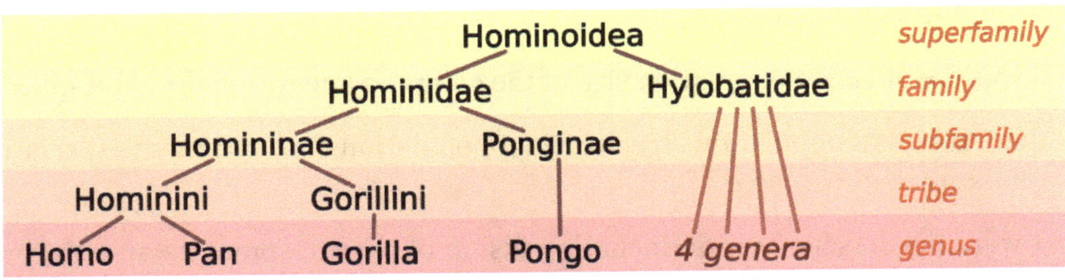

The taxonomic relationships of hominoids.

Biologists classify humans, along with only a few other species, as great apes (species in the family Hominidae). The living Hominidae include two distinct species of chimpanzee (the bonobo, *Pan paniscus*, and the common chimpanzee, *Pan troglodytes*), two species of gorilla (the western gorilla, *Gorilla gorilla*, and the eastern gorilla, *Gorilla graueri*), and two species of orangutan (the Bornean orangutan, *Pongo pygmaeus*, and the Sumatran orangutan, *Pongo abelii*). The great apes with the family Hylobatidae of gibbons form the superfamily Hominoidea of apes.

Apes, in turn, belong to the primate order (>400 species), along with the Old World monkeys, the New World monkeys, and others. Data from both mitochondrial DNA (mtDNA) and nuclear DNA (nDNA) indicate that primates belong to the group of Euarchontoglires, together with Rodentia, Lagomorpha, Dermoptera, and Scandentia. This is further supported by Alu-like short interspersed nuclear elements(SINEs) which have been found only in members of the Euarchontoglires.

Phylogenetic

A phylogenetic tree is usually derived from DNA or proteinsequences from populations. Often, mitochondrial DNA or Y chromosome sequences are used to study ancient human demographics. These single-locus sources of DNA do not recombine and are almost always inherited from a single parent, with only one known exception in mtDNA. Individuals from closer geographic regions generally tend to be more similar than individuals from regions farther away. Distance on a phylogenetic tree can be used approximately to indicate:

- Genetic distance. The genetic difference between humans and chimpanzees is less than 2%, or three times larger than the variation among modern humans (estimated at 0.6%).
- Temporal remoteness of the most recent common ancestor. The mitochondrial most recent common ancestor of modern humans is estimated to have lived roughly 160,000 years ago, the latest common ancestors of humans and chimpanzees roughly 5 to 6 million years ago.

Speciation of Humans and the African Apes

The separation of humans from their closest relatives, the non-human apes (chimpanzees and gorillas), has been studied extensively for more than a century. Five major questions have been addressed:

- Which apes are our closest ancestors?
- When did the separations occur?
- What was the effective population size of the common ancestor before the split?
- Are there traces of population structure (subpopulations) preceding the speciation or partial admixture succeeding it?
- What were the specific events (including fusion of chromosomes 2a and 2b) prior to and subsequent to the separation?

General Observations

The different parts of the genome show different sequence divergence between different hominoids. It has also been shown that the sequence divergence between DNA from humans and chimpanzees varies greatly. For example, the sequence divergence varies between 0% to 2.66% between non-coding, non-repetitive genomic regions of humans and chimpanzees. The percentage of nucleotides in the human genome (hg38) that had one-to-one exact matches in the chimpanzee genome (pantro6) was 84.38%. Additionally gene trees, generated by comparative analysis of DNA segments, do not always fit the species tree. Summing up:

- The sequence divergence varies significantly between humans, chimpanzees and gorillas.
- For most DNA sequences, humans and chimpanzees appear to be most closely related, but some point to a human-gorilla or chimpanzee-gorilla clade.
- The human genome has been sequenced, as well as the chimpanzee genome. Humans have 23 pairs of chromosomes, while chimpanzees, gorillas, and orangutans have 24. Human chromosome 2 is a fusion of two chromosomes 2a and 2b that remained separate in the other primates.

Divergence Times

The divergence time of humans from other apes is of great interest. One of the first molecular studies, published in 1967 measured immunological distances (IDs) between different primates. Basically the study measured the strength of immunological response that an antigen from one species (human albumin) induces in the immune system of another species (human, chimpanzee, gorilla and Old World monkeys). Closely related species should have similar antigens and therefore weaker immunological response to each other's antigens. The immunological response of a species to its own antigens (e.g. human to human) was set to be 1.

The ID between humans and gorillas was determined to be 1.09, that between humans and chimpanzees was determined as 1.14. However the distance to six different Old World monkeys was on average 2.46, indicating that the African apes are more closely related to humans than to monkeys. The authors consider the divergence time between Old World monkeys and hominoids to be 30 million years ago (MYA), based on fossil data, and the immunological distance was considered to grow at a constant rate. They concluded that divergence time of humans and the African apes to be roughly ~5 MYA. That was a surprising result. Most scientists at that time thought that humans and great apes diverged much earlier (>15 MYA).

The gorilla was, in ID terms, closer to human than to chimpanzees; however, the difference was so slight that the trichotomy could not be resolved with certainty. Later studies based on molecular genetics were able to resolve the trichotomy: chimpanzees are phylogenetically closer to humans than to gorillas. However, some divergence times estimated later (using much more sophisticated methods in molecular genetics) do not substantially differ from the very first estimate in 1967, but a recent paper puts it at 11–14 MYA.

Divergence Times and Ancestral Effective Population Size

Current methods to determine divergence times use DNA sequence alignments and molecular clocks. Usually the molecular clock is calibrated assuming that the orangutan split from the African apes (including humans) 12-16 MYA. Some studies also include some old world monkeys and set the divergence time of them from hominoids to 25-30 MYA. Both calibration points are based on very little fossil data and have been criticized.

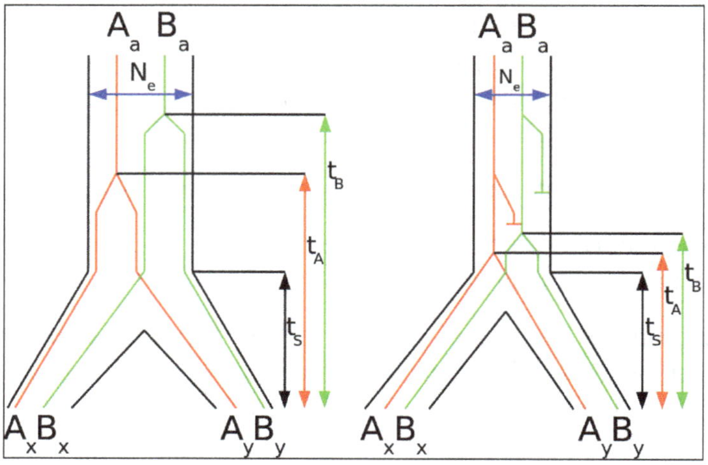

In above figure, the sequences of the DNA segments diverge earlier than the species. A large effective population size in the ancestral population (left) preserves different variants of the DNA segments (=alleles) for a longer period of time. Therefore, on average, the gene divergence times (t_A for DNA segment A; t_B for DNA segment B) will deviate more from the time the species diverge (t_S) compared to a small ancestral effective population size (right).

If these dates are revised, the divergence times estimated from molecular data will change as well. However, the relative divergence times are unlikely to change. Even if we can't tell absolute divergence times exactly, we can be pretty sure that the divergence time between chimpanzees and humans is about sixfold shorter than between chimpanzees (or humans) and monkeys.

One study used 15 DNA sequences from different regions of the genome from human and chimpanzee and 7 DNA sequences from human, chimpanzee and gorilla. They determined that chimpanzees are more closely related to humans than gorillas. Using various statistical methods, they estimated the divergence time human-chimp to be 4.7 MYA and the divergence time between gorillas and humans (and chimps) to be 7.2 MYA.

Additionally they estimated the effective population size of the common ancestor of humans and chimpanzees to be ~100,000. This was somewhat surprising since the present day effective population size of humans is estimated to be only ~10,000. If true that means that the human lineage would have experienced an immense decrease of its effective population size (and thus genetic diversity) in its evolution.

Another study sequenced 53 non-repetitive, intergenic DNA segments from a human, a chimpanzee, a gorilla, and orangutan. When the DNA sequences were concatenated to a single long sequence, the generated neighbor-joining tree supported the *Homo-Pan* clade with 100% bootstrap (that is that humans and chimpanzees are the closest related species of the four). When three species are fairly closely related to each other (like human, chimpanzee and gorilla), the trees obtained from DNA sequence data may not be congruent with the tree that represents the speciation (species tree).

The shorter internodal time span (T_{IN}) the more common are incongruent gene trees. The effective population size (N_e) of the internodal population determines how long genetic lineages are preserved in the population. A higher effective population size causes more incongruent gene trees. Therefore, if the internodal time span is known, the ancestral effective population size of the common ancestor of humans and chimpanzees can be calculated.

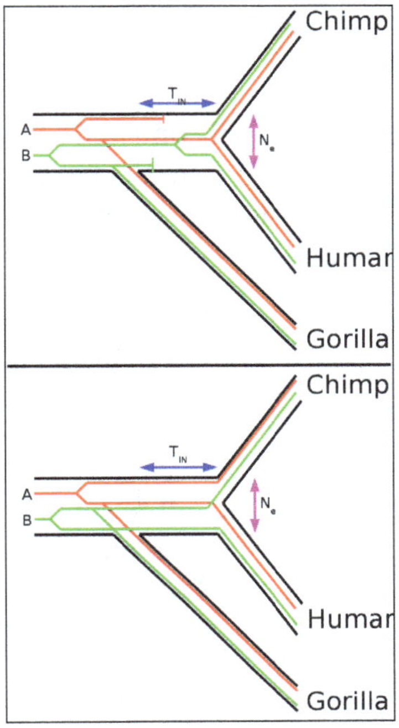

In above figure, A and B are two different loci. In the upper figure they fit to the species tree. The DNA that is present in today's gorillas diverged earlier from the DNA that is present in today's humans and chimps. Thus both loci should be more similar between human and chimp than between gorilla and chimp or gorilla and human. In the lower graph, locus A has a more recent common ancestor in human and gorilla compared to the chimp sequence. Whereas chimp and gorilla have a more recent common ancestor for locus B. Here the gene trees are incongruent to the species tree.

When each segment was analyzed individually, 31 supported the *Homo-Pan* clade, 10 supported the *Homo-Gorilla* clade, and 12 supported the *Pan-Gorilla* clade. Using the molecular clock the authors estimated that gorillas split up first 6.2-8.4 MYA and chimpanzees and humans split up 1.6-2.2 million years later (internodal time span) 4.6-6.2 MYA. The internodal time span is useful to estimate the ancestral effective population size of the common ancestor of humans and chimpanzees.

A parsimonious analysis revealed that 24 loci supported the *Homo-Pan* clade, 7 supported the *Homo-Gorilla* clade, 2 supported the *Pan-Gorilla* clade and 20 gave no resolution. Additionally they took 35 protein coding loci from databases. Of these 12 supported the *Homo-Pan*clade, 3 the *Homo-Gorilla* clade, 4 the *Pan-Gorilla* clade and 16 gave no resolution. Therefore, only ~70% of the 52 loci that gave a resolution (33 intergenic, 19 protein coding) support the 'correct' species tree. From the fraction of loci which did not support the species tree and the internodal time span they estimated previously, the effective population of the common ancestor of humans and chimpanzees was estimated to be ~52 000 to 96 000. This value is not as high as that from the first study (Takahata), but still much higher than present day effective population size of humans.

A third study used the same dataset that Chen and Li used but estimated the ancestral effective population of 'only' ~12,000 to 21,000, using a different statistical method.

Genetic Differences between Humans and other Great Apes

The alignable sequences within genomes of humans and chimpanzees differ by about 35 million single-nucleotide substitutions. Additionally about 3% of the complete genomes differ by deletions, insertions and duplications.

Since mutation rate is relatively constant, roughly one half of these changes occurred in the human lineage. Only a very tiny fraction of those fixed differences gave rise to the different phenotypes of humans and chimpanzees and finding those is a great challenge. The vast majority of the differences are neutral and do not affect the phenotype.

Molecular evolution may act in different ways, through protein evolution, gene loss, differential gene regulation and RNA evolution. All are thought to have played some part in human evolution.

Gene Loss

Many different mutations can inactivate a gene, but few will change its function in a specific way. Inactivation mutations will therefore be readily available for selection to act on. Gene loss could thus be a common mechanism of evolutionary adaptation (the "less-is-more" hypothesis).

80 genes were lost in the human lineage after separation from the last common ancestor with the chimpanzee. 36 of those were for olfactory receptors. Genes involved in chemoreception and immune response are overrepresented. Another study estimated that 86 genes had been lost.

Hair Keratin Gene KRTHAP1

A gene for type 1 hair keratin was lost in the human lineage. Keratins are a major component of hairs. Humans still have nine functional type I hair keratin genes, but the loss of that particular gene may have caused the thinning of human body hair. Based on the assumption of a constant molecular clock, the study predicts the gene loss occurred relatively recently in human evolution—less than 240 000 years ago, but both the Vindija Neandertal and the high-coverage Denisovan sequence contain the same premature stop codons as modern humans and hence dating should be greater than 750 000 years ago.

Myosin Gene MYH16

Stedman *et al.* stated that the loss of the sarcomeric myosin gene MYH16 in the human lineage led to smaller masticatory muscles. They estimated that the mutation that led to the inactivation (a two base pair deletion) occurred 2.4 million years ago, predating the appearance of *Homo ergaster/ erectus* in Africa. The period that followed was marked by a strong increase in cranial capacity, promoting speculation that the loss of the gene may have removed an evolutionary constraint on brain size in the genus *Homo*.

Another estimate for the loss of the MYH16 gene is 5.3 million years ago, long before *Homo* appeared.

Other

- CASPASE12, a cysteinyl aspartate proteinase. The loss of this gene is speculated to have reduced the lethality of bacterial infection in humans.

Gene Addition

Segmental duplications (SDs or LCRs) have had roles in creating new primate genes and shaping human genetic variation.

Human-specific DNA Insertions

When the human genome was compared to the genomes of five comparison primate species, including the chimpanzee, gorilla, orangutan, gibbon, and macaque, it was found that there are approximately 20,000 human-specific insertions believed to be regulatory. While most insertions appear to be fitness neutral, a small amount have been identified in positively selected genes showing associations to neural phenotypes and some relating to dental and sensory perception-related phenotypes. These findings hint at the seemingly important role of human-specific insertions in the recent evolution of humans.

Selection Pressures

Human accelerated regions are areas of the genome that differ between humans and chimpanzees to a greater extent than can be explained by genetic drift over the time since the two species shared a common ancestor. These regions show signs of being subject to natural selection, leading to the evolution of distinctly human traits. Two examples are HAR1F, which is believed to be related to brain development and HAR2 (a.k.a. HACNS1) that may have played a role in the development of the opposable thumb.

It has also been hypothesized that much of the difference between humans and chimpanzees is attributable to the regulation of gene expression rather than differences in the genes themselves. Analyses of conserved non-coding sequences, which often contain functional and thus positively selected regulatory regions, address this possibility.

Sequence Divergence between Humans and Apes

When the draft sequence of the common chimpanzee (*Pan troglodytes*) genome was published in the summer 2005, 2400 million bases (of ~3160 million bases) were sequenced and assembled well enough to be compared to the human genome. 1.23% of this sequenced differed by single-base substitutions. Of this, 1.06% or less was thought to represent fixed differences between the species, with the rest being variant sites in humans or chimpanzees. Another type of difference, called indels (insertions/deletions) accounted for many fewer differences (15% as many), but contributed ~1.5% of unique sequence to each genome, since each insertion or deletion can involve anywhere from one base to millions of bases.

A companion paper examined segmental duplications in the two genomes, whose insertion and deletion into the genome account for much of the indel sequence. They found that a total of 2.7% of euchromatic sequence had been differentially duplicated in one or the other lineage.

Percentage sequence divergence between humans and other hominids			
Locus	Human-Chimp	Human-Gorilla	Human-Orangutan
Alu elements	2	-	-
Non-coding (Chr. Y)	1.68 ± 0.19	2.33 ± 0.2	5.63 ± 0.35
Pseudogenes (autosomal)	1.64 ± 0.10	1.87 ± 0.11	-
Pseudogenes (Chr. X)	1.47 ± 0.17	-	-
Noncoding (autosomal)	1.24 ± 0.07	1.62 ± 0.08	3.08 ± 0.11
Genes (K_s)	1.11	1.48	2.98
Introns	0.93 ± 0.08	1.23 ± 0.09	-
Xq13.3	0.92 ± 0.10	1.42 ± 0.12	3.00 ± 0.18
Subtotal for X chromosome	1.16 ± 0.07	1.47 ± 0.08	-
Genes (K_a)	0.8	0.93	1.96

The sequence divergence has generally the following pattern: Human-Chimp < Human-Gorilla << Human-Orangutan, highlighting the close kinship between humans and the African apes. Alu elements diverge quickly due to their high frequency of CpG dinucleotides which mutate roughly 10 times more often than the average nucleotide in the genome. The mutation rate is higher in the male germ line, therefore the divergence in the Y chromosome—which is inherited solely from the father—is higher than in autosomes. The X chromosome is inherited twice as often through the female germ line as through the male germ line and therefore shows slightly lower sequence divergence. The sequence divergence of the Xq13.3 region is surprisingly low between humans and chimpanzees.

Mutations altering the amino acid sequence of proteins (K_a) are the least common. In fact ~29% of all orthologous proteins are identical between human and chimpanzee. The typical protein differs by only two amino acids. The measures of sequence divergence shown in the table only take the substitutional differences, for example from an A (adenine) to a G (guanine), into account. DNA sequences may however also differ by insertions and deletions (indels) of bases. These are usually stripped from the alignments before the calculation of sequence divergence is performed.

Genetic Differences between Modern Humans and Neanderthals

An international group of scientists completed a draft sequence of the Neanderthal genome in May 2010. The results indicate some breeding between modern humans (*Homo sapiens*) and Neanderthals (*Homo neanderthalensis*), as the genomes of non-African humans have 1–4% more in common with Neanderthals than do the genomes of subsaharan Africans. Neanderthals and most modern humans share a lactose-intolerantvariant of the lactase gene that encodes an enzyme that is unable to break down lactose in milk after weaning. Modern humans and Neanderthals also share the FOXP2 gene variant associated with brain development and with speech in modern humans, indicating that Neanderthals may have been able to speak. Chimps have two amino acid differences in FOXP2 compared with human and Neanderthal FOXP2.

Genetic Differences among Modern Humans

H. sapiens is thought to have emerged about 300,000 years ago. It dispersed throughout Africa, and after 70,000 years ago throughout Eurasia and Oceania. A 2009 study identified 14 "ancestral population clusters", the most remote being the San people of Southern Africa.

With their rapid expansion throughout different climate zones, and especially with the availability of new food sources with the domestication of cattle and the development of agriculture, human populations have been exposed to significant selective pressures since their dispersal. For example, East Asians have been found to be separated from Europids by a number of concentrated alleles suggestive of selection pressures, including variants of the EDAR, ADH1B, ABCC1, and ALDH-2genes. The East Asian types of ADH1B in particular are associated with rice domestication and would thus have arisen after the development of rice cultivation roughly 10,000 years ago. Several phenotypical traits of characteristic of East Asians are due to a single mutation of the EDAR gene, dated to c. 35,000 years ago.

As of 2017, the Single Nucleotide Polymorphism Database (dbSNP), which lists SNP and other variants, listed a total of 324 million variants found in sequenced human genomes. Nucleotide diversity, the average proportion of nucleotides that differ between two individuals, is estimated at between 0.1% and 0.4% for contemporary humans (compared to 2% between humans and chimpanzees). This corresponds to genome differences at a few million sites; the 1000 Genomes Project similarly found that "a typical [individual] genome differs from the reference human genome at 4.1 million to 5.0 million sites affecting 20 million bases of sequence."

In February 2019, scientists discovered evidence, based on genetics studies using artificial intelligence (AI), that suggest the existence of an unknown human ancestor species, not Neanderthal, Denisovan or human hybrid (like Denny (hybrid hominin)), in the genome of modern humans.

Human Mitochondrial Genetics

Human mitochondrial genetics is the study of the genetics of human mitochondrial DNA (the DNA contained in human mitochondria). The human mitochondrial genome is the entirety of hereditary information contained in human mitochondria. Mitochondria are small structures in cells that generate energy for the cell to use, and are hence referred to as the "powerhouses" of the cell.

Mitochondrial DNA (mtDNA) is not transmitted through nuclear DNA (nDNA). In humans, as in most multicellular organisms, mitochondrial DNA is inherited only from the mother's ovum. There are theories, however, that paternal mtDNA transmission in humans can occur under certain circumstances.

Mitochondrial inheritance is therefore non-Mendelian, as Mendelian inheritance presumes that half the genetic material of a fertilized egg (zygote) derives from each parent.

Eighty percent of mitochondrial DNA codes for mitochondrial RNA, and therefore most mitochondrial DNA mutations lead to functional problems, which may be manifested as muscle disorders (myopathies).

Because they provide 30 molecules of ATP per glucose molecule in contrast to the 2 ATP molecules produced by glycolysis, mitochondria are essential to all higher organisms for sustaining life. The mitochondrial diseases are genetic disorders carried in mitochondrial DNA, or nuclear DNA coding for mitochondrial components. Slight problems with any one of the numerous enzymes used by the mitochondria can be devastating to the cell, and in turn, to the organism.

Quantity

In humans, mitochondrial DNA (mtDNA) forms closed circular molecules that contain 16,569 DNA base pairs, with each such molecule normally containing a full set of the mitochondrial genes. Each human mitochondrion contains, on average, approximately 5 such mtDNA molecules, with the quantity ranging between 1 and 15. Each human cell contains approximately 100 mitochondria, giving a total number of mtDNA molecules per human cell of approximately 500.

Inheritance Patterns

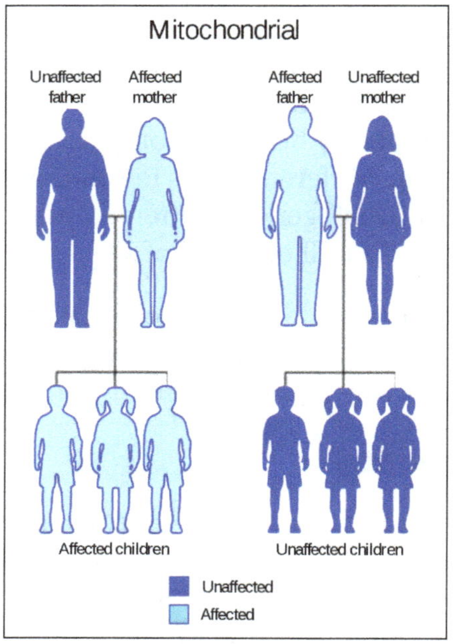

Mitochondrial inheritance patterns.

Because mitochondrial diseases (diseases due to malfunction of mitochondria) can be inherited both maternally and through chromosomal inheritance, the way in which they are passed on from generation to generation can vary greatly depending on the disease. Mitochondrial genetic mutations that occur in the nuclear DNA can occur in any of the chromosomes (depending on the species). Mutations inherited through the chromosomes can be autosomal dominant or recessive and can also be sex-linked dominant or recessive. Chromosomal inheritance follows normal Mendelian laws, despite the fact that the phenotype of the disease may be masked.

Because of the complex ways in which mitochondrial and nuclear DNA "communicate" and interact, even seemingly simple inheritance is hard to diagnose. A mutation in chromosomal DNA may change a protein that regulates (increases or decreases) the production of another certain protein in the mitochondria or the cytoplasm; this may lead to slight, if any, noticeable symptoms. On the other hand, some devastating mtDNA mutations are easy to diagnose because of their widespread damage to muscular, neural, and/or hepatic tissues (among other high-energy and metabolism-dependent tissues) and because they are present in the mother and all the offspring.

The number of affected mtDNA molecules inherited by a specific offspring can vary greatly because:

- The mitochondria within the fertilized oocyte is what the new life will have to begin with (in terms of mtDNA).
- The number of affected mitochondria varies from cell (in this case, the fertilized oocyte) to cell depending both on the number it inherited from its mother cell and environmental factors which may favor mutant or wildtypemitochondrial DNA.
- The number of mtDNA molecules in the mitochondria varies from around two to ten.

It is possible, even in twin births, for one baby to receive more than half mutant mtDNA molecules while the other twin may receive only a tiny fraction of mutant mtDNA molecules with respect to wildtype (depending on how the twins divide from each other and how many mutant mitochondria happen to be on each side of the division). In a few cases, some mitochondria or a mitochondrion from the sperm cell enters the oocyte but paternal mitochondria are actively decomposed.

Genes

Genes in the human mitochondrial genome are as follows:

Electron Transport Chain and Human

It was originally incorrectly believed that the mitochondrial genome contained only 13 protein-coding genes, all of them encoding proteins of the electron transport chain. However, in 2001, a 14th biologically active protein called humanin was discovered, and was found to be encoded by the mitochondrial gene MT-RNR2 which also encodes part of the mitochondrial ribosome (made out of RNA):

Complex number	Category	Genes	Positions in the mitogenome	Strand
I	NADH dehydrogenase			
		MT-ND1	3,307–4,262	L
		MT-ND2	4,470–5,511	L
		MT-ND3	10,059–10,404	L
		MT-ND4L	10,470–10,766	L
		MT-ND4	10,760–12,137 (overlap with MT-ND4L)	L
		MT-ND5	12,337–14,148	L
		MT-ND6	14,149–14,673	H
III	Coenzyme Q - cytochrome c reductase / Cytochrome b	MT-CYB	14,747–15,887	L
IV	Cytochrome c oxidase	MT-CO1	5,904–7,445	L
		MT-CO2	7,586–8,269	L
		MT-CO3	9,207–9,990	L
V	ATP synthase	MT-ATP6	8,527–9,207 (overlap with MT-ATP8)	L
		MT-ATP8	8,366–8,572	L
—	Humanin	MT-RNR2	—	—

Unlike the other proteins, humanin does not remain in the mitochondria, and interacts with the rest of the cell and cellular receptors. Humanin can protect brain cells by inhibiting apoptosis. Despite its name, versions of humanin also exist in other animals, such as rattin in rats.

rRNA

The following genes encode rRNAs:

Subunit	rRNA	Genes	Positions in the mitogenome	Strand
Small (SSU)	12S	MT-RNR1	648–1,601	L
Large (LSU)	16S	MT-RNR2	1,671–3,229	L

tRNA

The following genes encode tRNAs:

Amino Acid	3-Letter	1-Letter	MT DNA	Positions	Strand
Alanine	Ala	A	MT-TA	5,587–5,655	H
Arginine	Arg	R	MT-TR	10,405–10,469	L
Asparagine	Asn	N	MT-TN	5,657–5,729	H
Aspartic acid	Asp	D	MT-TD	7,518–7,585	L
Cysteine	Cys	C	MT-TC	5,761–5,826	H
Glutamic acid	Glu	E	MT-TE	14,674–14,742	H
Glutamine	Gln	Q	MT-TQ	4,329–4,400	H
Glycine	Gly	G	MT-TG	9,991–10,058	L
Histidine	His	H	MT-TH	12,138–12,206	L
Isoleucine	Ile	I	MT-TI	4,263–4,331	L
Leucine	Leu (UUR)	L	MT-TL1	3,230–3,304	L
Leucine	Leu (CUN)	L	MT-TL2	12,266–12,336	L
Lysine	Lys	K	MT-TK	8,295–8,364	L
Methionine	Met	M	MT-TM	4,402–4,469	L
Phenylalanine	Phe	F	MT-TF	577–647	L
Proline	Pro	P	MT-TP	15,956–16,023	H
Serine	Ser (UCN)	S	MT-TS1	7,446–7,514	H
Serine	Ser (AGY)	S	MT-TS2	12,207–12,265	L
Threonine	Thr	T	MT-TT	15,888–15,953	L
Tryptophan	Trp	W	MT-TW	5,512–5,579	L
Tyrosine	Tyr	Y	MT-TY	5,826–5,891	H
Valine	Val	V	MT-TV	1,602–1,670	L

Location of Genes

Mitochondrial DNA traditionally had the two strands of DNA designated the heavy and the light strand, due to their buoyant densities during separation in cesium chloride gradients, which was found to be related to the relative G+T nucleotide content of the strand. However, confusion of labeling of this strands is widespread, and appears to originate with a identification of the majority coding strand as the heavy. In humans, the light strand of mtDNA carries 28 genes and the heavy strand of mtDNA carries only 9 genes. Eight of the 9 genes on the heavy strand code for mitochondrial tRNA molecules. Human mtDNA consists of 16,569 nucleotide pairs. The entire molecule is regulated by only one regulatory region which contains the origins of replication of both heavy and light strands. The entire human mitochondrial DNA molecule has been mapped.

Genetic Code Variants

The genetic code is, for the most part, universal, with few exceptions: mitochondrial genetics includes some of these. For most organisms the "stop codons" are "UAA", "UAG", and "UGA". In vertebrate mitochondria "AGA" and "AGG" are also stop codons, but not "UGA", which codes for tryptophan instead. "AUA" codes for isoleucine in most organisms but for methionine in vertebrate mitochondrial mRNA.

There are many other variations among the codes used by other mitochondrial m/tRNA, which happened not to be harmful to their organisms, and which can be used as a tool (along with other mutations among the mtDNA/RNA of different species) to determine relative proximity of common ancestry of related species. (The more related two species are, the more mtDNA/RNA mutations will be the same in their mitochondrial genome).

Using these techniques, it is estimated that the first mitochondria arose around 1.5 billion years ago. A generally accepted hypothesis is that mitochondria originated as an aerobic prokaryote in a symbiotic relationship within an anaerobic eukaryote.

Replication, Repair, Transcription and Translation

Mitochondrial replication is controlled by nuclear genes and is specifically suited to make as many mitochondria as that particular cell needs at the time.

Mitochondrial transcription in humans is initiated from three promoters, H1, H2, and L (heavy strand 1, heavy strand 2, and light strand promoters). The H2 promoter transcribes almost the entire heavy strand and the L promoter transcribes the entire light strand. The H1 promoter causes the transcription of the two mitochondrial rRNA molecules.

When transcription takes place on the heavy strand a polycistronic transcript is created. The light strand produces either small transcripts, which can be used as primers, or one long transcript. The production of primers occurs by processing of light strand transcripts with the Mitochondrial RNase MRP (Mitochondrial RNA Processing). The requirement of transcription to produce primers links the process of transcription to mtDNA replication. Full length transcripts are cut into functional tRNA, rRNA, and mRNA molecules.

The process of transcription initiation in mitochondria involves three types of proteins: the mitochondrial RNA polymerase (POLRMT), mitochondrial transcription factor A (TFAM), and mitochondrial transcription factors B1 and B2 (TFB1M, TFB2M). POLRMT, TFAM, and TFB1M or TFB2M assemble at the mitochondrial promoters and begin transcription. The actual molecular events that are involved in initiation are unknown, but these factors make up the basal transcription machinery and have been shown to function in vitro.

Mitochondrial translation is still not very well understood. In vitro translations have still not been successful, probably due to the difficulty of isolating sufficient mt mRNA, functional mt rRNA, and possibly because of the complicated changes that the mRNA undergoes before it is translated.

Mitochondrial DNA Polymerase

The Mitochondrial DNA Polymerase (Pol gamma, encoded by the POLG gene) is used in the copying of mtDNA during replication. Because the two (heavy and light) strands on the circular mtDNA molecule have different origins of replication, it replicates in a D-loop mode. One strand begins to replicate first, displacing the other strand. This continues until replication reaches the origin of replication on the other strand, at which point the other strand begins replicating in the opposite direction. This results in two new mtDNA molecules. Each mitochondrion has several copies of the mtDNA molecule and the number of mtDNA molecules is a limiting factor in mitochondrial fission. After the mitochondrion has enough mtDNA, membrane area, and membrane proteins, it can undergo fission (very similar to that which bacteria use) to become two mitochondria. Evidence suggests that mitochondria can also undergo fusion and exchange (in a form of crossover) genetic material among each other. Mitochondria sometimes form large matrices in which fusion, fission, and protein exchanges are constantly occurring. mtDNA shared among mitochondria (despite the fact that they can undergo fusion).

Damage and Transcription Error

Mitochondrial DNA is susceptible to damage from free oxygen radicals from mistakes that occur during the production of ATP through the electron transport chain. These mistakes can be caused by genetic disorders, cancer, and temperature variations. These radicals can damage mtDNA molecules or change them, making it hard for mitochondrial polymerase to replicate them. Both cases can lead to deletions, rearrangements, and other mutations. Recent evidence has suggested that mitochondria have enzymes that proofread mtDNA and fix mutations that may occur due to free radicals. It is believed that a DNA recombinase found in mammalian cells is also involved in a repairing recombination process. Deletions and mutations due to free radicals have been associated with the aging process. It is believed that radicals cause mutations which lead to mutant proteins, which in turn led to more radicals. This process takes many years and is associated with some aging processes involved in oxygen-dependent tissues such as brain, heart, muscle, and kidney. Auto-enhancing processes such as these are possible causes of degenerative diseases including Parkinson's, Alzheimer's, and coronary artery disease.

Chromosomally Mediated mtDNA Replication Errors

Because mitochondrial growth and fission are mediated by the nuclear DNA, mutations in nuclear DNA can have a wide array of effects on mtDNA replication. Despite the fact that the loci for some of these mutations have been found on human chromosomes, specific genes and proteins involved have not yet been isolated. Mitochondria need a certain protein to undergo fission. If this protein (generated by the nucleus) is not present, the mitochondria grow but they do not divide. This leads to giant, inefficient mitochondria. Mistakes in chromosomal genes or their products can also affect mitochondrial replication more directly by inhibiting mitochondrial polymerase and can even cause mutations in the mtDNA directly and indirectly. Indirect mutations are most often caused by radicals created by defective proteins made from nuclear DNA.

Mitochondrial Diseases

Contribution of Mitochondrial versus Nuclear Genome

In total, the mitochondrion hosts about 3000 different types of proteins, but only about 13 of them are coded on the mitochondrial DNA. Most of the 3000 types of proteins are involved in a variety of processes other than ATP production, such as porphyrin synthesis. Only about 3% of them code for ATP production proteins. This means most of the genetic information coding for the protein makeup of mitochondria is in chromosomal DNA and is involved in processes other than ATP synthesis. This increases the chances that a mutation that will affect a mitochondrion will occur in chromosomal DNA, which is inherited in a Mendelian pattern. Another result is that a chromosomal mutation will affect a specific tissue due to its specific needs, whether those may be high energy requirements or a need for the catabolism or anabolism of a specific neurotransmitter or nucleic acid. Because several copies of the mitochondrial genome are carried by each mitochondrion (2–10 in humans), mitochondrial mutations can be inherited maternally by mtDNA mutations which are present in mitochondria inside the oocytebefore fertilization, or (as stated above) through mutations in the chromosomes.

Presentation

Mitochondrial diseases range in severity from asymptomatic to fatal, and are most commonly due to inherited rather than acquired mutations of mitochondrial DNA. A given mitochondrial mutation can cause various diseases depending on the severity of the problem in the mitochondria and the tissue the affected mitochondria are in. Conversely, several different mutations may present themselves as the same disease. This almost patient-specific characterization of mitochondrial diseases makes them very hard to accurately recognize, diagnose and trace. Some diseases are observable at or even before birth (many causing death) while others do not show themselves until late adulthood (late-onset disorders). This is because the number of mutant versus wildtype mitochondria varies between cells and tissues, and is continuously changing. Because cells have multiple mitochondria, different mitochondria in the same cell can have different variations of the mtDNA. This condition is referred to as heteroplasmy. When a certain tissue reaches a certain ratio of mutant versus wildtype mitochondria, a disease will present itself. The ratio varies from person to person and tissue to tissue (depending on its specific energy, oxygen, and metabolism requirements, and the effects of the specific mutation). Mitochondrial diseases are very numerous and different. Apart from diseases caused by abnormalities in mitochondrial DNA, many diseases are suspected to be associated in part by mitochondrial dysfunctions, such as diabetes mellitus, forms of cancer and cardiovascular disease, lactic acidosis, specific forms of myopathy, osteoporosis, Alzheimer's disease, Parkinsons's disease, stroke, male infertility and which are also believed to play a role in the aging process.

Use in Identification

Human mtDNA can also be used to help identify individuals. Forensic laboratories occasionally use mtDNA comparison to identify human remains, and especially to identify older unidentified skeletal remains. Although unlike nuclear DNA, mtDNA is not specific to one individual, it can be used in combination with other evidence (anthropological evidence, circumstantial evidence, and the like) to establish identification. mtDNA is also used to exclude possible matches between missing persons and unidentified remains. Many researchers believe that mtDNA is better suited to identification of older skeletal remains than nuclear DNA because the greater number of copies of mtDNA per cell increases the chance of obtaining a useful sample, and because a match with a living relative is possible even if numerous maternal generations separate the two. American outlaw Jesse James's remains were identified using a comparison between mtDNA extracted from his remains and the mtDNA of the son of the female-line great-granddaughter of his sister. Similarly, the remains of Alexandra Feodorovna (Alix of Hesse), last Empress of Russia, and her children were identified by comparison of their mitochondrial DNA with that of Prince Philip, Duke of Edinburgh, whose maternal grandmother was Alexandra's sister Victoria of Hesse. Similarly to identify Emperor Nicholas II remains his mitochondrial DNA was compared with that of James Carnegie, 3rd Duke of Fife, whose maternal great-grandmother Alexandra of Denmark (Queen Alexandra) was sister of Nicholas II mother Dagmar of Denmark (Empress Maria Feodorovna).

BEHAVIOR GENETICS

Behavior genetics, also called psychogenetics is the study of the influence of an organism's genetic composition on its behavior and the interaction of heredity and environment insofar as they affect behavior. The question of the determinants of behavioral abilities and disabilities has commonly been referred to as the "nature-nurture" controversy.

The relationship between behavior and genetics, or heredity, dates to the work of English scientist Sir Francis Galton, who coined the phrase "nature and nurture." Galton studied the families of outstanding men of his day and concluded, like his cousin Charles Darwin, that mental powers run in families. Galton became the first to use twins in genetic research and pioneered many of the statistical methods of analysis that are in use today. In 1918 British statistician and geneticist Ronald Aylmer Fisher published a paper that showed how Gregor Mendel's laws of inheritance applied to complex traits influenced by multiple genes and environmental factors.

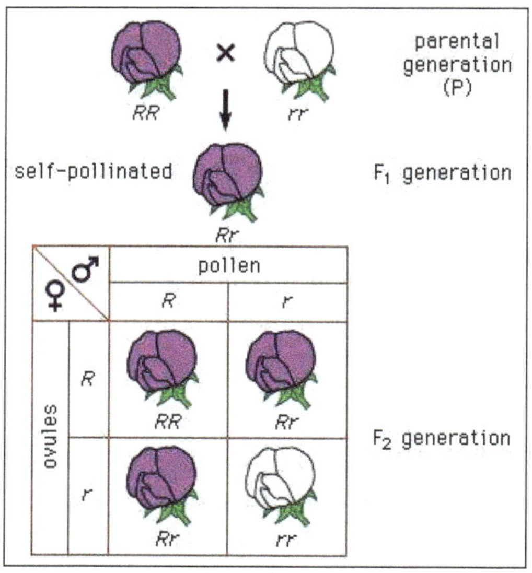

Mendel's law of segregationCross of a purple-flowered and a white-flowered strain of peas. R stands for the gene for purple flowers and r for the gene for white flowers.

The first human behavioral genetic research on intelligence and mental illness began in the 1920s, when environmentalism (the theory that behavior is a result of nongenetic factors such as various childhood experiences) became popular and before Nazi Germany's abuse of genetics made the notion of hereditary influence abhorrent. Although genetic research on human behavior continued throughout the following decades, it was not until the 1970s that a balanced view came to prevail in psychiatry that recognized the importance of nature as well as nurture. In psychology, this reconciliation did not take hold until the 1980s. Much behavioral genetic research today focuses on identifying specific genes that affect behavioral dimensions, such as personality and intelligence, and disorders, such as autism, hyperactivity, depression, and schizophrenia.

Methods of Study

Quantitative genetic methods are used to estimate the net effect of genetic and environmental factors on individual differences in any complex trait, including behavioral traits. In addition, molecular genetic methods are used to identify specific genes responsible for genetic influence. Research is carried out in both animals and humans; however, studies using animal models tend to provide more-accurate data than studies in humans because both genes and environment can be manipulated and controlled in the laboratory.

By mating related animals such as siblings for many generations, nearly pure strains are obtained in which all offspring are genetically highly similar. It is possible to screen for genetic influence on behavior by comparing the behavior of different inbred strains raised in the same laboratory environment. Another method, known as selective breeding, evaluates genetic involvement by attempting to breed for high and low extremes of a trait for several generations. Both methods have been applied to a wide variety of animal behaviors, especially learning and behavioral responses to drugs, and this research provides evidence for widespread influence of genes on behavior.

Because genes and environments cannot be manipulated in the human species, two quasi-experimental methods are used to screen for genetic influence on individual differences in complex traits such as behavior. The twin method relies on the accident of nature that results in identical (monozygotic, MZ) twins or fraternal (dizygotic, DZ) twins. MZ twins are like clones, genetically identical to each other because they came from the same fertilized egg. DZ twins, on the other hand, developed from two eggs that happened to be fertilized at the same time. Like other siblings, DZ twins are only half as similar genetically as MZ twins. To the extent that behavioral variability is caused by environmental factors, DZ twins should be as similar for the behavioral trait as are MZ twins because both types of twins are reared by the same parents in the same place at the same time. If the trait is influenced by genes, then DZ twins ought to be less similar than MZ twins. For schizophrenia, for example, the concordance (risk of one twin's being schizophrenic if the other is) is about 45 percent for MZ twins and about 15 percent for DZ twins. For intelligence as assessed by IQ tests, the correlation, an index of resemblance (0.00 indicates no resemblance and 1.00 indicates perfect resemblance), is 0.85 for MZ twins and 0.60 for DZ twins for studies throughout the world of more than 10,000 pairs of twins. The twin method has been robustly defended as a rough screen for genetic influence on behavior.

The adoption method is a quasi-experimental design that relies on a social accident in which children are adopted away from their biological (birth) parents early in life, thus cleaving the effects of nature and nurture. Because the twin and adoption methods are so different, greater confidence is warranted when results from these two methods converge on the same conclusion—as they usually do. An influential adoption study of schizophrenia in 1966 by American behavioral geneticist Leonard Heston showed that children adopted away from their schizophrenic biological mothers at birth were just as likely to become schizophrenic (about 10 percent) as were children reared by their schizophrenic biological mothers. A 20-year study begun in the 1970s in the United States of intelligence of adopted children and their biological and adoptive parents showed increasing similarity from infancy to childhood to adolescence between the adopted children and their biological parents but no resemblance between the adopted children and their adoptive parents.

In contrast to traditional molecular genetic research that focused on rare disorders caused by a single genetic mutation, molecular genetic research on complex behavioral traits and common behavioral disorders is much more difficult because multiple genes are involved and each gene has a relatively small effect. However, some genes identified in animal models have contributed to an improved understanding of complex human behavioral disorders such as reading disability, hyperactivity, autism, and dementia.

CLASSICAL GENETICS

Classical genetics is the branch of genetics based solely on visible results of reproductive acts. It is the oldest discipline in the field of genetics, going back to the experiments on Mendelian inheritance by Gregor Mendelwho made it possible to identify the basic mechanisms of heredity. Subsequently, these mechanisms have been studied and explained at the molecular level.

Classical genetics consists of the techniques and methodologies of genetics that were in use before the advent of molecular biology. A key discovery of classical genetics in eukaryotes was genetic linkage. The observation that some genes do not segregate independently at meiosis broke the laws of Mendelian inheritance, and provided science with a way to map characteristics to a location on the chromosomes. Linkage maps are still used today, especially in breeding for plant improvement.

After the discovery of the genetic code and such tools of cloning as restriction enzymes, the avenues of investigation open to geneticists were greatly broadened. Some classical genetic ideas have been supplanted with the mechanistic understanding brought by molecular discoveries, but many remain intact and in use. Classical genetics is often contrasted with reverse genetics, and aspects of molecular biology are sometimes referred to as molecular genetics.

At the base of classical genetics is the concept of a gene, the hereditary factor tied to a particular simple feature (or character).

The set of genes for one or more characters possessed by an individual is the genotype. A diploid individual often has two alleles for the determination of a character.

Classical genetics is the part of genetics that is solely about the method in which genetic traits are transmitted via the acts of reproduction. Genetics is, generally, the study of genes, genetic variation, and heredity. The process by which characteristics are passed down from parents to their offspring is called heredity. In the sense of classical genetics, variation is known as the lack of resemblance in related individuals and can be categorized as discontinuous or continuous. Genes are a fundamental part of DNA that is aligned linearly on a eukaryotic chromosome. Chemical information that is transported and encoded by each gene is referred to as a trait. Many organisms possess two genes for each individual trait that is present within that particular individual. These paired genes that control the same trait is classified as an allele. In an individual, the allelic genes that are expressed can be either homozygous, meaning the same, or heterozygous, meaning different. Many pairs of alleles have differing effects that are portrayed in an offspring's phenotype and genotype. The phenotype is a general term that defines an individual's visible, physical traits. The genotype of an offspring is known as its genetic makeup. The alleles of genes can either be dominant or recessive. A dominant allele needs only one copy to be expressed while a recessive allele needs two copies (homozygous) in a diploid organism to be expressed. Dominant and recessive alleles help to determine the offspring's genotypes, and therefore phenotypes.

Classical genetics is often referred to as the oldest form of genetics, and began with Gregor Mendel'sexperiments that formulated and defined a fundamental biological concept known as Mendelian Inheritance. Mendelian Inheritance is the process in which genes and traits are passed from a set of parents to their offspring. These inherited traits are passed down mechanistically with one gene from one parent and the second gene from another parent in sexually reproducing organisms. This creates the pair of genes in diploid organisms. Gregor Mendel started his experimentation and study of inheritance with phenotypes of garden peas and continued the experiments with plants. He focused on the patterns of the traits that were being passed down from one generation to the next generation. This was assessed by test-crossing two peas of different colors and observing the resulting phenotypes. After determining how the traits were likely inherited, he began to expand the amount of traits observed and tested and eventually expanded his experimentation by increasing the number of different organisms he tested.

About 150 years ago, Gregor Mendel published his first experiments with the test crossing of *Pisum* peas. Seven different phenotypic characteristics were studied and tested in the peas, including seed color, flower color and seed shape. Mendel took peas that had differing phenotypic characteristics and test-crossed them to assess how the parental plants passed the traits down to their offspring. He started by crossing a round, yellow and round, green pea and observed the resulting phenotypes. The results of this experiment allowed him to see which of these two traits was dominant and which was recessive based upon the number of offspring with each phenotype. Mendel then chose to further his experiments by crossing a pea plant homozygous dominant for round and yellow phenotypes with a pea plant that was homozygous recessive for wrinkled and green. The plants that were originally crossed are known as the parental generation, or P generation, and the offspring resulting from the parental cross is known as the first filial, or F1, generation. The plants of the F1 generation resulting from this hybrid cross were all heterozygous round and yellow seeds.

Classical genetics is a hallmark of the start of great discovery in biology, and has led to increased understanding of multiple important components of molecular genetics, human genetics, medical genetics, and much more. Thus, reinforcing Mendel's nickname as the father of modern genetics.

CYTOGENETICS

Cytogenetics, in cell biology, is the field that deals with chromosomes and their inheritance, particularly as applied to medical genetics. Chromosomes are microscopic structures found in cells, and malformations associated with them lead to numerous genetic diseases. Chromosomal analysis has steadily improved in precision and resolution, and that has led to improvements in the diagnosis of various genetic diseases in all areas of medicine.

Genetic diseases begin with a mutation in chromosomal structure. Among the hundreds of types of genetic diseases are congenital abnormalities (birth defects), reproductive wastage (fetal loss), and developmental delays. Approximately 3 percent of infants are born with a congenital abnormality, and 50 percent of all spontaneous abortions involve some form of chromosomal defect. Some common genetic disorders are Down syndrome, Turner syndrome, cystic fibrosis, and Huntington disease.

The study of chromosomes begins with the extraction of intact chromosomes from live cells. Chromosomal analyses often make use of white blood cells (T lymphocytes), which multiply quickly under cell culture conditions. Chromosomes may also be extracted from skin cells, bone marrow cells, or fetal cells (by amniocentesis or chorionic villus sampling).

The 23 pairs of chromosomes can be identified by using various staining techniques, such as Giemsa banding (G-banding), quinacrine banding (Q-banding), reverse banding (R-banding), constitutive heterochromatin (or centromere) banding (C-banding), and fluorescence in situ hybridization (FISH). G-banding is one of the most-used chromosomal staining methods. In this approach, chromosomes are first treated with an enzyme known as trypsin and then with Giemsa stain. All chromosomes can be individually identified by means of this technique. With Q-banding, chromosomes are stained with quinacrine mustard or a related compound and examined by fluorescence microscopy, which is useful for detecting heteromorphisms (variants in chromosomal structural). In chemical staining with the R-banding method, chromosomes are first treated with heat, resulting in patterns that are easier to analyze. In C-banding, regions of chromosomes containing heterochromatin (the condensed, inactive form of DNA) are stained.

FISH is used to examine specific DNA sequences and the number or structure of chromosomes. The technique is based on the use of fluorescent probes that are capable of detecting particular DNA sequences. FISH is a rapid and highly sensitive technique and often is used to detect genetic abnormalities in embryos (preimplantation genetic diagnosis).

MEDICAL GENETICS

Medical genetics is the branch of medicine that involves the diagnosis and management of hereditary disorders. Medical genetics differs from human genetics in that human genetics is a field of scientific research that may or may not apply to medicine, while medical genetics refers to the application of genetics to medical care. For example, research on the causes and inheritance of genetic disorders would be considered within both human genetics and medical genetics, while the diagnosis, management, and counselling people with genetic disorders would be considered part of medical genetics.

In contrast, the study of typically non-medical phenotypes such as the genetics of eye color would be considered part of human genetics, but not necessarily relevant to medical genetics (except in situations such as albinism). *Genetic medicine* is a newer term for medical genetics and incorporates areas such as gene therapy, personalized medicine, and the rapidly emerging new medical specialty, predictive medicine.

Scope

Medical genetics encompasses many different areas, including clinical practice of physicians, genetic counselors, and nutritionists, clinical diagnostic laboratory activities, and research into the causes and inheritance of genetic disorders. Examples of conditions that fall within the scope of medical genetics include birth defects and dysmorphology, mental retardation, autism, mitochondrial disorders, skeletal dysplasia, connective tissue disorders, cancer genetics, teratogens, and prenatal diagnosis. Medical genetics is increasingly becoming relevant to many common diseases. Overlaps with other medical specialties are beginning to emerge, as recent advances in genetics are revealing etiologies for neurologic, endocrine, cardiovascular, pulmonary, ophthalmologic, renal, psychiatric, and dermatologic conditions. The medical genetics community is increasingly involved with individuals who have undertaken elective genetic and genomic testing.

Subspecialties

In some ways, many of the individual fields within medical genetics are hybrids between clinical care and research. This is due in part to recent advances in science and technology that have enabled an unprecedented understanding of genetic disorders.

Clinical Genetics

Clinical genetics is the practice of clinical medicine with particular attention to hereditary disorders. Referrals are made to genetics clinics for a variety of reasons, including birth defects, developmental delay, autism, epilepsy, short stature, and many others. Examples of genetic syndromes that are commonly seen in the genetics clinic include chromosomal rearrangements, Down syndrome, DiGeorge syndrome, Fragile X syndrome, Marfan syndrome, Neurofibromatosis, Turner syndrome, and Williams syndrome.

In the United States, Doctors who practice clinical genetics are accredited by the American Board of Medical Genetics and Genomics (ABMGG). In order to become a board-certified practitioner of Clinical Genetics, a physician must complete a minimum of 24 months of training in a program accredited by the ABMGG. Individuals seeking acceptance into clinical genetics training programs must hold an M.D. or D.O. degree (or their equivalent) and have completed a minimum of 24 months of training in an ACGME-accredited residencyprogram in internal medicine, pediatrics, obstetrics and gynecology, or other medical specialty.

Metabolic/Biochemical Genetics

Metabolic (or biochemical) genetics involves the diagnosis and management of inborn errors of metabolism in which patients have enzymatic deficiencies that perturb biochemical pathways involved in metabolism of carbohydrates, amino acids, and lipids. Examples of metabolic disorders include galactosemia, glycogen storage disease, lysosomal storage disorders, metabolic acidosis, peroxisomal disorders, phenylketonuria, and urea cycle disorders.

Cytogenetics

Cytogenetics is the study of chromosomes and chromosome abnormalities. While cytogenetics historically relied on microscopy to analyze chromosomes, new molecular technologies such as array comparative genomic hybridization are now becoming widely used. Examples of chromosome abnormalities include aneuploidy, chromosomal rearrangements, and genomic deletion/duplication disorders.

Molecular Genetics

Molecular genetics involves the discovery of and laboratory testing for DNA mutations that underlie many single gene disorders. Examples of single gene disorders include achondroplasia, cystic fibrosis, Duchenne muscular dystrophy, hereditary breast cancer (BRCA1/2), Huntington disease, Marfan syndrome, Noonan syndrome, and Rett syndrome. Molecular tests are also used in the diagnosis of syndromes involving epigenetic abnormalities, such as Angelman syndrome, Beckwith-Wiedemann syndrome, Prader-willi syndrome, and uniparental disomy.

Mitochondrial Genetics

Mitochondrial genetics concerns the diagnosis and management of mitochondrial disorders, which have a molecular basis but often result in biochemical abnormalities due to deficient energy production.

There exists some overlap between medical genetic diagnostic laboratories and molecular pathology.

Genetic Counseling

Genetic counseling is the process of providing information about genetic conditions, diagnostic testing, and risks in other family members, within the framework of nondirective counseling. Genetic counselors are non-physician members of the medical genetics team who specialize in family risk assessment and counseling of patients regarding genetic disorders. The precise role of the genetic counselor varies somewhat depending on the disorder. When working alongside geneticists, genetic counselors normally specialize in pediatric genetics which focuses on developmental abnormalities present in newborns, infants or children. The major goal of pediatric counseling is attempting to explain the genetic basis behind the child›s developmental concerns in a compassionate and articulated manner that allows the potentially distressed or frustrated parents to easily understand the information. As well, genetic counselors normally take a family pedigree, which summarizes the medical history of the patient›s family. This then aids the clinical geneticist in the differential diagnosis process and help determine which further steps should be taken to help the patient.

Current Practice

The clinical setting in which patients are evaluated determines the scope of practice, diagnostic, and therapeutic interventions. For the purposes of general discussion, the typical encounters between patients and genetic practitioners may involve:

- Referral to an out-patient genetics clinic (pediatric, adult, or combined) or an in-hospital consultation, most often for diagnostic evaluation.
- Specialty genetics clinics focusing on management of inborn errors of metabolism, skeletal dysplasia, or lysosomal storage diseases.
- Referral for counseling in a prenatal genetics clinic to discuss risks to the pregnancy (advanced maternal age, teratogen exposure, family history of a genetic disease), test results (abnormal maternal serum screen, abnormal ultrasound), and/or options for prenatal diagnosis (typically non-invasive prenatal screening, diagnostic amniocentesis or chorionic villus sampling).
- Multidisciplinary specialty clinics that include a clinical geneticist or genetic counselor (cancer genetics, cardiovascular genetics, craniofacial or cleft lip/palate, hearing loss clinics, muscular dystrophy/neurodegenerative disorder clinics).

Diagnostic Evaluation

Each patient will undergo a diagnostic evaluation tailored to their own particular presenting signs and symptoms. The geneticist will establish a differential diagnosis and recommend appropriate testing. These tests might evaluate for chromosomal disorders, inborn errors of metabolism, or single gene disorders.

Chromosome Studies

Chromosome studies are used in the general genetics clinic to determine a cause for developmental delay/mental retardation, birth defects, dysmorphic features, and/or autism. Chromosome analysis is also performed in the prenatal setting to determine whether a fetus is affected with aneuploidy or other chromosome rearrangements. Finally, chromosome abnormalities are often detected in cancer samples. A large number of different methods have been developed for chromosome analysis:

- Chromosome analysis using a karyotype involves special stains that generate light and dark bands, allowing identification of each chromosome under a microscope.
- Fluorescence in situ hybridization (FISH) involves fluorescent labeling of probes that bind to specific DNA sequences, used for identifying aneuploidy, genomic deletions or duplications, characterizing chromosomal translocations and determining the origin of ring chromosomes.
- Chromosome painting is a technique that uses fluorescent probes specific for each chromosome to differentially label each chromosome. This technique is more often used in cancer cytogenetics, where complex chromosome rearrangements can occur.
- Array comparative genomic hybridization is a newer molecular technique that involves hybridization of an individual DNA sample to a glass slide or microarray chip containing molecular probes (ranging from large ~200kb bacterial artificial chromosomes to small

oligonucleotides) that represent unique regions of the genome. This method is particularly sensitive for detection of genomic gains or losses across the genome but does not detect balanced translocations or distinguish the location of duplicated genetic material (for example, a tandem duplication versus an insertional duplication).

Basic Metabolic Studies

Biochemical studies are performed to screen for imbalances of metabolites in the bodily fluid, usually the blood (plasma/serum) or urine, but also in cerebrospinal fluid (CSF). Specific tests of enzyme function (either in leukocytes, skin fibroblasts, liver, or muscle) are also employed under certain circumstances. In the US, the newborn screen incorporates biochemical tests to screen for treatable conditions such as galactosemia and phenylketonuria (PKU). Patients suspected to have a metabolic condition might undergo the following tests:

- Quantitative amino acid analysis is typically performed using the ninhydrin reaction, followed by liquid chromatography to measure the amount of amino acid in the sample (either urine, plasma/serum, or CSF). Measurement of amino acids in plasma or serum is used in the evaluation of disorders of amino acid metabolism such as urea cycle disorders, maple syrup urine disease, and PKU. Measurement of amino acids in urine can be useful in the diagnosis of cystinuria or renal Fanconi syndrome as can be seen in cystinosis.
- Urine organic acid analysis can be either performed using quantitative or qualitative methods, but in either case the test is used to detect the excretion of abnormal organic acids. These compounds are normally produced during bodily metabolism of amino acids and odd-chain fatty acids, but accumulate in patients with certain metabolic conditions.
- The acylcarnitine combination profile detects compounds such as organic acids and fatty acids conjugated to carnitine. The test is used for detection of disorders involving fatty acid metabolism, including MCAD.
- Pyruvate and lactate are byproducts of normal metabolism, particularly during anaerobic metabolism. These compounds normally accumulate during exercise or ischemia, but are also elevated in patients with disorders of pyruvate metabolism or mitochondrial disorders.
- Ammonia is an end product of amino acid metabolism and is converted in the liver to urea through a series of enzymatic reactions termed the urea cycle. Elevated ammonia can therefore be detected in patients with urea cycle disorders, as well as other conditions involving liver failure.
- Enzyme testing is performed for a wide range of metabolic disorders to confirm a diagnosis suspected based on screening tests.

Molecular Studies

- DNA sequencing is used to directly analyze the genomic DNA sequence of a particular gene. In general, only the parts of the gene that code for the expressed protein (exons) and small amounts of the flanking untranslated regions and introns are analyzed. Therefore, although these tests are highly specific and sensitive, they do not routinely identify all of the mutations that could cause disease.

- DNA methylation analysis is used to diagnose certain genetic disorders that are caused by disruptions of epigenetic mechanisms such as genomic imprinting and uniparental disomy.
- Southern blotting is an early technique basic on detection of fragments of DNA separated by size through gel electrophoresis and detected using radiolabeled probes. This test was routinely used to detect deletions or duplications in conditions such as Duchenne muscular dystrophy but is being replaced by high-resolution array comparative genomic hybridization techniques. Southern blotting is still useful in the diagnosis of disorders caused by trinucleotide repeats.

Treatments

Each cell of the body contains the hereditary information (DNA) wrapped up in structures called chromosomes. Since genetic syndromes are typically the result of alterations of the chromosomes or genes, there is no treatment currently available that can correct the genetic alterations in every cell of the body. Therefore, there is currently no "cure" for genetic disorders. However, for many genetic syndromes there is treatment available to manage the symptoms. In some cases, particularly inborn errors of metabolism, the mechanism of disease is well understood and offers the potential for dietary and medical management to prevent or reduce the long-term complications. In other cases, infusion therapy is used to replace the missing enzyme. Current research is actively seeking to use gene therapy or other new medications to treat specific genetic disorders.

Management of Metabolic Disorders

In general, metabolic disorders arise from enzyme deficiencies that disrupt normal metabolic pathways. For instance, in the hypothetical example:

```
A ---> B ---> C ---> D      AAAA ---> BBBBBB ---> CCCCCCCCCC ---> no D)
  X       Y      Z               X           Y        |     (no or insufficient Z)
                                                    EEEEE
```

Compound "A" is metabolized to "B" by enzyme "X", compound "B" is metabolized to "C" by enzyme "Y", and compound "C" is metabolized to "D" by enzyme "Z". If enzyme "Z" is missing, compound "D" will be missing, while compounds "A", "B", and "C" will build up. The pathogenesis of this particular condition could result from lack of compound "D", if it is critical for some cellular function, or from toxicity due to excess "A", "B", and/or "C", or from toxicity due to the excess of "E" which is normally only present in small amounts and only accumulates when "C" is in excess. Treatment of the metabolic disorder could be achieved through dietary supplementation of compound "D" and dietary restriction of compounds "A", "B", and/or "C" or by treatment with a medication that promoted disposal of excess "A", "B", "C" or "E". Another approach that can be taken is enzyme replacement therapy, in which a patient is given an infusion of the missing enzyme "Z" or cofactor therapy to increase the efficacy of any residual "Z" activity.

- Diet

Dietary restriction and supplementation are key measures taken in several well-known metabolic disorders, including galactosemia, phenylketonuria (PKU), maple syrup urine disease, organic acidurias and urea cycle disorders. Such restrictive diets can be difficult for the patient and family to maintain, and require close consultation with a nutritionist who has special experience in metabolic disorders. The composition of the diet will change depending on the caloric needs of the growing child and special attention is needed during a pregnancy if a woman is affected with one of these disorders.

- Medication

Medical approaches include enhancement of residual enzyme activity (in cases where the enzyme is made but is not functioning properly), inhibition of other enzymes in the biochemical pathway to prevent buildup of a toxic compound, or diversion of a toxic compound to another form that can be excreted. Examples include the use of high doses of pyridoxine (vitamin B6) in some patients with homocystinuria to boost the activity of the residual cystathione synthase enzyme, administration of biotin to restore activity of several enzymes affected by deficiency of biotinidase, treatment with NTBC in Tyrosinemia to inhibit the production of succinylacetone which causes liver toxicity, and the use of sodium benzoate to decrease ammonia build-up in urea cycle disorders.

- Enzyme replacement therapy

Certain lysosomal storage diseases are treated with infusions of a recombinant enzyme (produced in a laboratory), which can reduce the accumulation of the compounds in various tissues. Examples include Gaucher disease, Fabry disease, Mucopolysaccharidoses and Glycogen storage disease type II. Such treatments are limited by the ability of the enzyme to reach the affected areas (the blood brain barrier prevents enzyme from reaching the brain, for example), and can sometimes be associated with allergic reactions. The long-term clinical effectiveness of enzyme replacement therapies vary widely among different disorders.

Other Examples

- Angiotensin receptor blockers in Marfan syndrome & Loeys-Dietz.
- Bone marrow transplantation.
- Gene therapy.

Ethical, Legal and Social Implications

Genetic information provides a unique type of knowledge about an individual and his/her family, fundamentally different from a typically laboratory test that provides a "snapshot" of an individual's health status. The unique status of genetic information and inherited disease has a number of ramifications with regard to ethical, legal, and societal concerns.

On 19 March 2015, scientists urged a worldwide ban on clinical use of methods, particularly the use of CRISPRand zinc finger, to edit the human genome in a way that can be inherited. In April 2015 and April 2016, Chinese researchers reported results of basic research to edit the DNA of non-viable human embryos using CRISPR. In February 2016, British scientists were given permission by regulators to genetically modify human embryos by using CRISPR and related techniques on condition that the embryos were destroyed within seven days. In June 2016 the Dutch government was reported to be planning to follow suit with similar regulations which would specify a 14-day limit.

Societies

The more empirical approach to human and medical genetics was formalized by the founding in 1948 of the American Society of Human Genetics. The Society first began annual meetings that year and its international counterpart, the International Congress of Human Genetics, has met every 5 years since its inception in 1956.

Medical genetics is now recognized as a distinct medical specialty in the U.S. with its own approved board (the American Board of Medical Genetics) and clinical specialty college (the American College of Medical Genetics). The College holds an annual scientific meeting, publishes a monthly journal, *Genetics in Medicine*, and issues position papers and clinical practice guidelines on a variety of topics relevant to human genetics.

Research

The broad range of research in medical genetics reflects the overall scope of this field, including basic research on genetic inheritance and the human genome, mechanisms of genetic and metabolic disorders, translational research on new treatment modalities, and the impact of genetic testing.

Basic Genetics Research

Basic research geneticists usually undertake research in universities, biotechnology firms and research institutes.

Allelic Architecture of Disease

Sometimes the link between a disease and an unusual gene variant is more subtle. The genetic architecture of common diseases is an important factor in determining the extent to which patterns of genetic variation influence group differences in health outcomes. According to the common disease/common variant hypothesis, common variants present in the ancestral population before the dispersal of modern humans from Africa play an important role in human diseases. Genetic variants associated with Alzheimer disease, deep venous thrombosis, Crohn disease, and type 2 diabetes appear to adhere to this model. However, the generality of the model has not yet been established and, in some cases, is in doubt. Some diseases, such as many common cancers, appear not to be well described by the common disease/common variant model.

Another possibility is that common diseases arise in part through the action of combinations of variants that are individually rare. Most of the disease-associated alleles discovered to date have been rare, and rare variants are more likely than common variants to be differentially distributed among groups distinguished by ancestry. However, groups could harbor different, though perhaps overlapping, sets of rare variants, which would reduce contrasts between groups in the incidence of the disease.

The number of variants contributing to a disease and the interactions among those variants also could influence the distribution of diseases among groups. The difficulty that has been encountered in finding contributory alleles for complex diseases and in replicating positive associations suggests that many complex diseases involve numerous variants rather than a moderate number of alleles, and the influence of any given variant may depend in critical ways on the genetic and environmental background. If many alleles are required to increase susceptibility to a disease, the odds are low that the necessary combination of alleles would become concentrated in a particular group purely through drift.

Population Substructure in Genetics Research

One area in which population categories can be important considerations in genetics research is in controlling for confounding between population substructure, environmental exposures, and health outcomes. Association studies can produce spurious results if cases and controls have differing allele frequencies for genes that are not related to the disease being studied, although the magnitude of this problem in genetic association studies is subject to debate. Various methods have been developed to detect and account for population substructure, but these methods can be difficult to apply in practice.

Population substructure also can be used to advantage in genetic association studies. For example, populations that represent recent mixtures of geographically separated ancestral groups can exhibit longer-range linkage disequilibrium between susceptibility alleles and genetic markers than is the case for other populations. Genetic studies can use this admixture linkage disequilibrium to search for disease alleles with fewer markers than would be needed otherwise. Association studies also can take advantage of the contrasting experiences of racial or ethnic groups, including migrant groups, to search for interactions between particular alleles and environmental factors that might influence health.

MOLECULAR GENETICS

Molecular genetics is the field of biology that studies the structure and function of genes at a molecular level and thus employs methods of both molecular biology and genetics. The study of chromosomes and gene expression of an organism can give insight into heredity, genetic variation, and mutations. This is useful in the study of developmental biology and in understanding and treating genetic diseases.

Technique

Amplification

Gene amplification is a procedure in which a certain gene or DNA sequence is replicated many times in a process called DNA replication.

Polymerase Chain Reaction

The main genetic components of the polymerase chain reaction (PCR) are DNA nucleotides, template DNA, primers and Taq polymerase. DNA nucleotides make up the DNA template strand for the specific sequence being amplified and primers are short strands of complementary nucleotides where DNA replication starts. Taq polymerase is a heat stable enzyme that jump-starts the production of new DNA at the high temperatures needed for reaction.

Cloning DNA in Bacteria

Cloning is the process of creating many identical copies of a sequence of DNA. The target DNA sequence is inserted into a cloning vector. Because this vector originates from a self-replicating virus, plasmid, or higher organism cell, when the appropriate size DNA is inserted, the "target and vector DNA fragments are then ligated" to create a recombinant DNA molecule.

The recombinant DNA molecule is then inserted into a bacterial strain (usually *E. coli*) which produces several identical copies of the selected sequence it absorbed through transformation (the mechanism by which bacteria uptake foreign DNA from the environment into their genomes).

Separation and Detection

In separation and detection, DNA and mRNA are isolated from cells and then detected simply by the isolation. Cell cultures are also grown to provide a constant supply of cells ready for isolation.

Cell Cultures

A cell culture for molecular genetics is a culture that is grown in artificial conditions. Some cell types grow well in cultures such as skin cells, but other cells are not as productive in cultures. There are different techniques for each type of cell, some only recently being found to foster growth in stem and nerve cells. Cultures for molecular genetics are frozen in order to preserve all copies of the gene specimen and thawed only when needed. This allows for a steady supply of cells.

DNA Isolation

DNA isolation extracts DNA from a cell in a pure form. First, the DNA is separated from cellular components such as proteins, RNA, and lipids. This is done by placing the chosen cells in a tube with a solution that mechanically, chemically, breaks the cells open. This solution contains enzymes, chemicals, and salts that breaks down the cells except for the DNA. It contains enzymes to dissolve proteins, chemicals to destroy all RNA present, and salts to help pull DNA out of the solution. Next, the DNA is separated from the solution by being spun in a centrifuge, which allows the DNA to collect in the bottom of the tube. After this cycle in the centrifuge the solution is poured off and the DNA is resuspended in a second solution that makes the DNA easy to work with in the future. This results in a concentrated DNA sample that contains thousands of copies of each gene. For large scale projects such as sequencing the human genome, all this work is done by robots.

mRNA Isolation

Expressed DNA that codes for the synthesis of a protein is the final goal for scientists and this expressed DNA is obtained by isolating mRNA (Messenger RNA).

First, laboratories use a normal cellular modification of mRNA that adds up to 200 adenine nucleotides to the end of the molecule (poly(A) tail). Once this has been added, the cell is ruptured and its cell contents are exposed to synthetic beads that are coated with thymine string nucleotides. Because Adenine and Thymine pair together in DNA, the poly(A) tail and synthetic beads are attracted to one another, and once they bind in this process the cell components can be washed away without removing the mRNA. Once the mRNA has been isolated, reverse transcriptase is employed to convert it to single-stranded DNA, from which a stable double-stranded DNA is produced using DNA polymerase. Complementary DNA (cDNA) is much more stable than mRNA and so, once the double-stranded DNA has been produced it represents the expressed DNA sequence scientists look for.

Genetic Screens

Forward Genetics

This technique is used to identify which genes or genetic mutations produce a certain phenotype. A mutagen is very often used to accelerate this process. Once mutants have been isolated, the mutated genes can be molecularly identified.

Forward saturation genetics is a method for treating organisms with a mutagen, then screens the organism's offspring for particular phenotypes. This type of genetic screening is used to find and identify all the genes involved in a trait.

Reverse Genetics

Reverse genetics determines the phenotype that results from a specifically engineered gene. In some organisms, such as yeast and mice, it is possible to induce the deletion of a particular gene, creating what's known as a gene "knockout" - the laboratory origin of so-called "knockout mice" for further study. In other words this process involves the creation of transgenic organisms that do not express a gene of interest. Alternative methods of reverse genetic research include the random induction of DNA deletions and subsequent selection for deletions in a gene of interest, as well as the application of RNA interference.

Gene Therapy

A mutation in a gene can cause encoded proteins and the cells that rely on those proteins to malfunction. Conditions related to gene mutations are called genetic disorders. However, altering a patient's genes can sometimes be used to treat or cure a disease as well. Gene therapy can be used to replace a mutated gene with the correct copy of the gene, to inactivate or knockout the expression of a malfunctioning gene, or to introduce a foreign gene to the body to help fight disease. Major diseases that can be treated with gene therapy include viral infections, cancers, and inherited disorders, including immune system disorders.

Gene therapy delivers a copy of the missing, mutated, or desired gene via a modified virus or vector to the patient's target cells so that a functional form of the protein can then be produced and incorporated into the body. These vectors are often siRNA. Treatment can be either in vivo or ex vivo. The therapy has to be repeated several times for the infected patient to continually be relieved, as repeated cell division and cell death slowly reduces the body's ratio of functional-to-mutant genes. Gene therapy is an appealing alternative to some drug-based approaches, because gene therapy repairs the underlying genetic defect using the patients own cells with minimal side effects.

Gene therapies are still in development and mostly used in research settings. All experiments and products are controlled by the U.S. FDA and the NIH.

Classical gene therapies usually require efficient transfer of cloned genes into the disease cells so that the introduced genes are expressed at sufficiently high levels to change the patient's physiology. There are several different physicochemical and biological methods that can be used to transfer genes into human cells. The size of the DNA fragments that can be transferred is very limited, and often the transferred gene is not a conventional gene. Horizontal gene transfer is the transfer of genetic material from one cell to another that is not its offspring. Artificial horizontal gene transfer is a form of genetic engineering.

POPULATION GENETICS

Population genetics is a subfield of genetics that deals with genetic differences within and between populations, and is a part of evolutionary biology. Studies in this branch of biology examine such phenomena as adaptation, speciation, and population structure.

Population genetics was a vital ingredient in the emergence of the modern evolutionary synthesis. Its primary founders were Sewall Wright, J. B. S. Haldane and Ronald Fisher, who also laid the foundations for the related discipline of quantitative genetics. Traditionally a highly mathematical discipline, modern population genetics encompasses theoretical, lab, and field work. Population genetic models are used both for statistical inference from DNA sequence data and for proof/disproof of concept.

What sets population genetics apart today from newer, more phenotypic approaches to modelling evolution, such as evolutionary game theory and adaptive dynamics, is its emphasis on genetic phenomena as dominance, epistasis, and the degree to which genetic recombination breaks up linkage disequilibrium. This makes it appropriate for comparison to population genomics data.

Four Processes

Selection

Natural selection, which includes sexual selection, is the fact that some traits make it more likely for an organismto survive and reproduce. Population genetics describes natural selection by defining fitness as a propensity or probability of survival and reproduction in a particular environment. The fitness is normally given by the symbol $w=1-s$ where s is the selection coefficient. Natural selection acts on phenotypes, so population genetic models assume relatively simple relationships to predict the phenotype and hence fitness from the allele at one or a small number of loci. In this way, natural selection converts differences in the fitness of individuals with different phenotypes into changes in allele frequency in a population over successive generations.

Before the advent of population genetics, many biologists doubted that small differences in fitness were sufficient to make a large difference to evolution. Population geneticists addressed this concern in part by comparing selection to genetic drift. Selection can overcome genetic drift when s is greater than 1 divided by the effective population size. When this criterion is met, the probability that a new advantageous mutant becomes fixed is approximately equal to 2s. The time until fixation of such an allele depends little on genetic drift, and is approximately proportional to log(sN)/s.

Dominance

Dominance means that the phenotypic and/or fitness effect of one allele at a locus depends on which allele is present in the second copy for that locus. Consider three genotypes at one locus, with the following fitness values:

-	Genotype:	A1A1	A1A2	A2A2	-	Relative fitness:	1	1-hs	1-s

The s is selection coefficient and h is the dominance coefficient. The value of h yields the following information:

-	h=0	A_1 dominant, A_2 recessive	-	h=1	A_2 dominant, A_1 recessive	-	$0<h<1$	incomplete dominance	-	$h<0$	overdominance	-	$h>1$	Underdominance

Epistasis

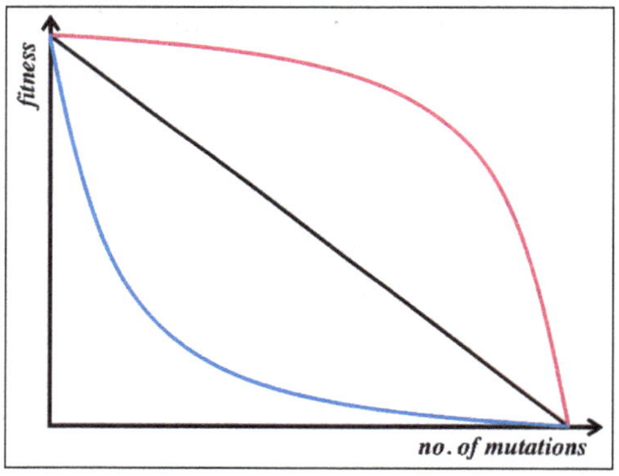

In the above figure, logarithm of fitness as a function of the number of deleterious mutations. Synergistic epistasis is represented by the red line - each subsequent deleterious mutation has a larger proportionate effect on the organism's fitness. Antagonistic epistasis is in blue. The black line shows the non-epistatic case, where fitness is the product of the contributions from each of its loci.

Epistasis means that the phenotypic and/or fitness effect of an allele at one locus depends on which alleles are present at other loci. Selection does not act on a single locus, but on a phenotype that arises through development from a complete genotype. However, many population genetics models of sexual species are "single locus" models, where the fitness of an individual is calculated as the product of the contributions from each of its loci—effectively assuming no epistasis.

In fact, the genotype to fitness landscape is more complex. Population genetics must either model this complexity in detail, or capture it by some simpler average rule. Empirically, beneficial mutations tend to have a smaller fitness benefit when added to a genetic background that already has high fitness: this is known as diminishing returns epistasis. When deleterious mutations also have a smaller fitness effect on high fitness backgrounds, this is known as "synergistic epistasis". However, the effect of deleterious mutations tends on average to be very close to multiplicative, or can even show the opposite pattern, known as "antagonistic epistasis".

Synergistic epistasis is central to some theories of the purging of mutation load and to the evolution of sexual reproduction.

Mutation

Drosophila melanogaster.

Mutation is the ultimate source of genetic variation in the form of new alleles. In addition, mutation may influence the direction of evolution when there is mutation bias, i.e. different probabilities for different mutations to occur. For example, recurrent mutation that tends to be in the opposite direction to selection can lead to mutation-selection balance. At the molecular level, if mutation from G to A happens more often than mutation from A to G, then genotypes with A will tend to evolve. Different insertion vs. deletion mutation biases in different taxa can lead to the evolution of different genome sizes. Developmental or mutational biases have also been observed in morphological evolution. For example, according to the phenotype-first theory of evolution, mutations can eventually cause the genetic assimilation of traits that were previously induced by the environment.

Mutation bias effects are superimposed on other processes. If selection would favor either one out of two mutations, but there is no extra advantage to having both, then the mutation that occurs the most frequently is the one that is most likely to become fixed in a population.

Mutation can have no effect, alter the product of a gene, or prevent the gene from functioning. Studies in the fly *Drosophila melanogaster* suggest that if a mutation changes a protein produced by a gene, this will probably be harmful, with about 70 percent of these mutations having damaging effects, and the remainder being either neutral or weakly beneficial. Most loss of function mutations are selected against. But when selection is weak, mutation bias towards loss of function can affect evolution. For example, pigments are no longer useful when animals live in the darkness of caves, and tend to be lost. This kind of loss of function can occur because of mutation bias, and/or because the function had a cost, and once the benefit of the function disappeared, natural selection leads to the loss. Loss of sporulation ability in a bacterium during laboratory evolution appears to have been caused by mutation bias, rather than natural selection against the cost of maintaining sporulation ability.When there is no selection for loss of function, the speed at which loss evolves depends more on the mutation rate than it does on the effective population size, indicating that it is driven more by mutation bias than by genetic drift.

Mutations can involve large sections of DNA becoming duplicated, usually through genetic recombination.This leads to copy-number variation within a population. Duplications are a major source of raw material for evolving new genes. Other types of mutation occasionally create new genes from previously noncoding DNA.

Genetic Drift

Genetic drift is a change in allele frequencies caused by random sampling. That is, the alleles in the offspring are a random sample of those in the parents. Genetic drift may cause gene variants to disappear completely, and thereby reduce genetic variability. In contrast to natural selection, which makes gene variants more common or less common depending on their reproductive success, the changes due to genetic drift are not driven by environmental or adaptive pressures, and are equally likely to make an allele more common as less common.

The effect of genetic drift is larger for alleles present in few copies than when an allele is present in many copies. The population genetics of genetic drift are described using either branching processes or a diffusion equationdescribing changes in allele frequency. These approaches are usually applied to the Wright-Fisher and Moranmodels of population genetics. Assuming genetic drift is the only evolutionary force acting on an allele, after t generations in many replicated populations, starting with allele frequencies of p and q, the variance in allele frequency across those populations is:

$$V_t \approx pq\left(1-\exp\left\{-\frac{t}{2N_e}\right\}\right).$$

Ronald Fisher held the view that genetic drift plays at the most a minor role in evolution, and this remained the dominant view for several decades. No population genetics perspective have ever given genetic drift a central role by itself, but some have made genetic drift important in combination with another non-selective force. The shifting balance theory of Sewall Wright held that the combination of population structure and genetic drift was important. Motoo Kimura's neutral theory of molecular evolution claims that most genetic differences within and between populations are caused by the combination of neutral mutations and genetic drift.

The role of genetic drift by means of sampling error in evolution has been criticized by John H Gillespie and Will Provine, who argue that selection on linked sites is a more important stochastic force, doing the work traditionally ascribed to genetic drift by means of sampling error. The mathematical properties of genetic draft are different from those of genetic drift. The direction of the random change in allele frequency is autocorrelated across generations.

Gene Flow

Because of physical barriers to migration, along with the limited tendency for individuals to move or spread (vagility), and tendency to remain or come back to natal place (philopatry), natural populations rarely all interbreed as may be assumed in theoretical random models (panmixy). There is usually a geographic range within which individuals are more closely related to one another than those randomly selected from the general population. This is described as the extent to which a population is genetically structured. Genetic structuring can be caused by migration due to historical climate change, species range expansion or current availability of habitat. Gene flow is hindered by mountain ranges, oceans and deserts or even man-made structures such as the Great Wall of China, which has hindered the flow of plant genes.

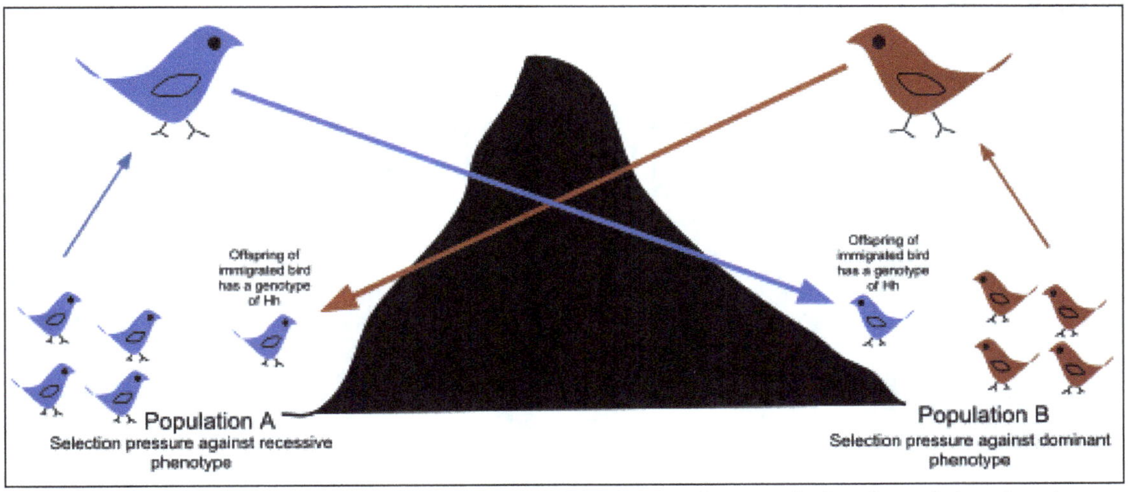

Gene flow is the transfer of alleles from one population to another population through immigration of individuals. In this example, one of the birds from population A immigrates to population B, which has fewer of the dominant alleles, and through mating incorporates its alleles into the other population.

The Great Wall of China is an obstacle to gene flow of some terrestrial species.

Gene flow is the exchange of genes between populations or species, breaking down the structure. Examples of gene flow within a species include the migration and then breeding of organisms, or the exchange of pollen. Gene transfer between species includes the formation of hybrid organisms and horizontal gene transfer. Population genetic models can be used to identify which populations show significant genetic isolation from one another, and to reconstruct their history.

Subjecting a population to isolation leads to inbreeding depression. Migration into a population can introduce new genetic variants, potentially contributing to evolutionary rescue. If a significant proportion of individuals or gametes migrate, it can also change allele frequencies, e.g. giving rise to migration load.

In the presence of gene flow, other barriers to hybridization between two diverging populations of an outcrossingspecies are required for the populations to become new species.

Horizontal Gene Transfer

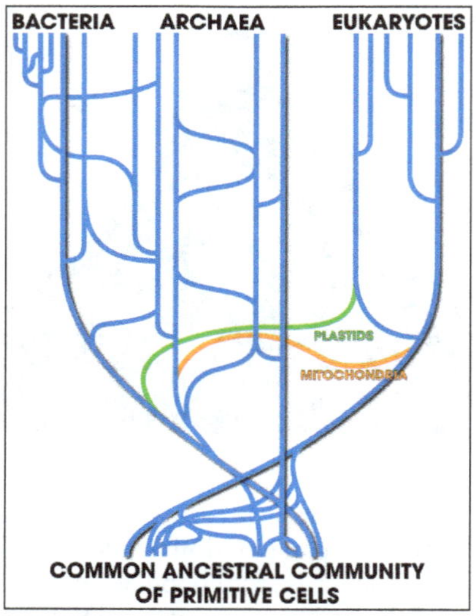

Current tree of life showing vertical and horizontal gene transfers.

Horizontal gene transfer is the transfer of genetic material from one organism to another organism that is not its offspring; this is most common among prokaryotes. In medicine, this contributes to the spread of antibiotic resistance, as when one bacteria acquires resistance genes it can rapidly transfer them to other species.Horizontal transfer of genes from bacteria to eukaryotes such as the yeast *Saccharomyces cerevisiae* and the adzuki bean beetle *Callosobruchus chinensis* may also have occurred. An example of larger-scale transfers are the eukaryotic bdelloid rotifers, which appear to have received a range of genes from bacteria, fungi, and plants. Viruses can also carry DNA between organisms, allowing transfer of genes even across biological domains. Large-scale gene transfer has also occurred between the ancestors of eukaryotic cells and prokaryotes, during the acquisition of chloroplasts and mitochondria.

Linkage

If all genes are in linkage equilibrium, the effect of an allele at one locus can be averaged across the gene poolat other loci. In reality, one allele is frequently found in linkage disequilibrium with genes at other loci, especially with genes located nearby on the same chromosome. Recombination breaks up this linkage disequilibrium too slowly to avoid genetic hitchhiking, where an allele at one locus rises to high frequency because it is linked to an allele under selection at a nearby locus. Linkage also slows down the rate of adaptation, even in sexual populations. The effect of linkage disequilibrium in slowing down the rate of adaptive evolution arises from a combination of the Hill–Robertson effect (delays in bringing beneficial mutations together) and background selection (delays in separating beneficial mutations from deleterious hitchhikers).

Linkage is a problem for population genetic models that treat one gene locus at a time. It can, however, be exploited as a method for detecting the action of natural selection via selective sweeps.

In the extreme case of an asexual population, linkage is complete, and population genetic equations can be derived and solved in terms of a travelling wave of genotype frequencies along a simple fitness landscape. Most microbes, such as bacteria, are asexual. The population genetics of their adaptation have two contrasting regimes. When the product of the beneficial mutation rate and population size is small, asexual populations follow a "successional regime" of origin-fixation dynamics, with adaptation rate strongly dependent on this product. When the product is much larger, asexual populations follow a "concurrent mutations" regime with adaptation rate less dependent on the product, characterized by clonal interference and the appearance of a new beneficial mutation before the last one has fixed.

Applications

Explaining Levels of Genetic Variation

Neutral theory predicts that the level of nucleotide diversity in a population will be proportional to the product of the population size and the neutral mutation rate. The fact that levels of genetic diversity vary much less than population sizes do is known as the "paradox of variation". While high levels of genetic diversity were one of the original arguments in favor of neutral theory, the paradox of variation has been one of the strongest arguments against neutral theory.

It is clear that levels of genetic diversity vary greatly within a species as a function of local recombination rate, due to both genetic hitchhiking and background selection. Most current solutions to the paradox of variation invoke some level of selection at linked sites. For example, one analysis suggests that larger populations have more selective sweeps, which remove more neutral genetic diversity. A negative correlation between mutation rate and population size may also contribute.

Life history affects genetic diversity more than population history does, e.g. r-strategists have more genetic diversity.

Detecting Selection

Population genetics models are used to infer which genes are undergoing selection. One common approach is to look for regions of high linkage disequilibrium and low genetic variance along the chromosome, to detect recent selective sweeps.

A second common approach is the McDonald–Kreitman test. The McDonald–Kreitman test compares the amount of variation within a species (polymorphism) to the divergence between species (substitutions) at two types of sites, one assumed to be neutral. Typically, synonymous sites are assumed to be neutral. Genes undergoing positive selection have an excess of divergent sites relative to polymorphic sites. The test can also be use to obtain a genome-wide estimate of the proportion of substitutions that are fixed by positive selection, α. According to the neutral theory of molecular evolution, this number should be near zero. High numbers have therefore been interpreted as a genome-wide falsification of neutral theory.

Demographic Inference

The simplest test for population structure in a sexually reproducing, diploid species, is to see whether genotype frequencies follow Hardy-Weinberg proportions as a function of allele frequencies.

For example, in the simplest case of a single locus with two alleles denoted A and a at frequencies p and q, random mating predicts freq(AA) = p^2 for the AA homozygotes, freq(aa) = q^2 for the aa homozygotes, and freq(Aa) = $2pq$ for the heterozygotes. In the absence of population structure, Hardy-Weinberg proportions are reached within 1-2 generations of random mating. More typically, there is an excess of homozygotes, indicative of population structure. The extent of this excess can be quantified as the inbreeding coefficient, F.

Individuals can be clustered into K subpopulations. The degree of population structure can then be calculated using F_{ST}, which is a measure of the proportion of genetic variance that can be explained by population structure. Genetic population structure can then be related to geographic structure, and genetic admixture can be detected.

Coalescent theory relates genetic diversity in a sample to demographic history of the population from which it was taken. It normally assumes neutrality, and so sequences from more neutrally-evolving portions of genomes are therefore selected for such analyses. It can be used to infer the relationships between species (phylogenetics), as well as the population structure, demographic history (e.g. population bottlenecks, population growth), biological dispersal, source-sink dynamics and introgression within a species.

Another approach to demographic inference relies on the allele frequency spectrum.

Evolution of Genetic Systems

By assuming that there are loci that control the genetic system itself, population genetic models are created to describe the evolution of dominance and other forms of robustness, the evolution of sexual reproduction and recombination rates, the evolution of mutation rates, the evolution of evolutionary capacitors, the evolution of costly signalling traits, the evolution of ageing, and the evolution of co-operation. For example, most mutations are deleterious, so the optimal mutation rate for a species may be a trade-off between the damage from a high deleterious mutation rate and the metabolic costs of maintaining systems to reduce the mutation rate, such as DNA repair enzymes.

One important aspect of such models is that selection is only strong enough to purge deleterious mutations and hence overpower mutational bias towards degradation if the selection coefficient s is greater than the inverse of the effective population size. This is known as the drift barrier and is related to the nearly neutral theory of molecular evolution. Drift barrier theory predicts that species with large effective population sizes will have highly streamlined, efficient genetic systems, while those with small population sizes will have bloated and complex genomes containing for example introns and transposable elements. However, somewhat paradoxically, species with large population sizes might be so tolerant to the consequences of certain types of errors that they evolve higher error rates, e.g. in transcription and translation, than small populations.

PLANT GENETICS

Plant genetics is the study of genes, genetic variation, and heredityspecifically in Plants. It is generally considered a field of biology and botany, but intersects frequently with many other life sciences and is strongly linked with the study of information systems. Plant genetics is similar in many ways to animal genetics but differs in a few key areas.

An image of multiple chromosomes, taken from many cells.

The discoverer of genetics was Gregor Mendel, a late 19th-century scientist and Augustinian friar. Mendel studied "trait inheritance", patterns in the way traits are handed down from parents to offspring. He observed that organisms (most famously pea plants) inherit traits by way of discrete "units of inheritance". This term, still used today, is a somewhat ambiguous definition of what is referred to as a gene. Much of Mendel's work with plants still forms the basis for modern plant genetics.

Plants, like all known organisms, use DNA to pass on their traits. Animal genetics often focuses on parentage and lineage, but this can sometimes be difficult in plant genetics due to the fact that plants can, unlike most animals, be self-fertile. Speciation can be easier in many plants due to unique genetic abilities, such as being well adapted to polyploidy. Plants are unique in that they are able to produce energy-dense carbohydrates via photosynthesis, a process which is achieved by use of Chloroplast|chloroplasts. Chloroplasts, like the superficially similar mitochondria, possess their own DNA. Chloroplasts thus provide an additional reservoir for genes and genetic diversity, and an extra layer of genetic complexity not found in animals.

The study of plant genetics has major economic impacts: many staple crops are genetically modified to increase yields, confer pest and disease resistance, provide resistance to herbicides, or to increase their nutritional value.

DNA

Deoxyribonucleic acid (DNA) is a nucleic acid that contains the genetic instructions used in the development and functioning of all known living organisms and some viruses. The main role of DNA molecules is the long-term storage of information. DNA is often compared to a set of blueprints or a recipe, or a code, since it contains the instructions needed to construct other components of cells, such as proteins and RNA molecules. The DNA segments that carry this genetic information are called genes, and their location within the genome are referred to as genetic loci, but other DNA sequences have structural purposes, or are involved in regulating the use of this genetic information.

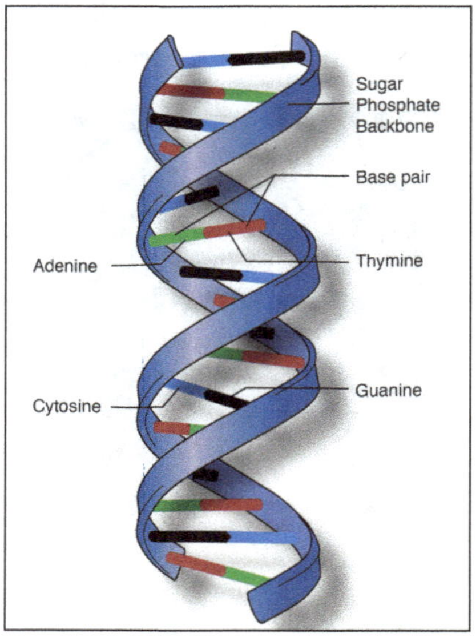

The structure of part of a DNA double helix.

Geneticists, including plant geneticists, use this sequence of DNA to their advantage to better find and understand the role of different genes within a given genome. Through research and plant breeding, manipulation of different plant genes and loci encoded by the DNA sequence of the plant chromosomes by various methods can be done to produce different or desired genotypes that result in different or desired phenotypes.

Plant Specific Genetics

Plants, like all other known living organisms, pass on their traits using DNA. Plants however are unique from other living organisms in the fact that they have Chloroplasts. Like mitochondria, chloroplasts have their own DNA. Like animals, plants experience somatic mutations regularly, but these mutations can contribute to the germ line with ease, since flowers develop at the ends of branches composed of somatic cells. People have known of this for centuries, and mutant branches are called "sports". If the fruit on the sport is economically desirable, a new cultivar may be obtained.

Some plant species are capable of self-fertilization, and some are nearly exclusively self-fertilizers. This means that a plant can be both mother and father to its offspring, a rare occurrence in animals. Scientists and hobbyists attempting to make crosses between different plants must take special measures to prevent the plants from self-fertilizing. In plant breeding, people create hybrids between plant species for economic and aesthetic reasons. For example, the yield of Corn has increased nearly five-fold in the past century due in part to the discovery and proliferation of hybrid corn varieties. Plant genetics can be used to predict which combination of plants may produce a plant with Hybrid vigor, or conversely many discoveries in Plant genetics have come from studying the effects of hybridization.

Plants are generally more capable of surviving, and indeed flourishing, as polyploids. Polyploid organisms have more than two sets of homologous chromosomes. For example, humans have two sets of homologous chromosomes, meaning that a typical human will have 2 copies each of 23 different chromosomes, for a total of 46. Wheat on the other hand, while having only 7 distinct chromosomes, is considered a hexaploid and has 6 copies of each chromosome, for a total of 42. In animals, inheritable germline polyploidy is less common, and spontaneous chromosome increases may not even survive past fertilization. In plants however this is no such problem, polyploid individuals are created frequently by a variety of processes, however once created usually cannot cross back to the parental type. Polyploid individuals, if capable of self-fertilizing, can give rise to a new genetically distinct lineage, which can be the start of a new species. This is often called "instant speciation". Polyploids generally have larger fruit, an economically desirable trait, and many human food crops, including wheat, maize, potatoes, peanuts, strawberries and tobacco, are either accidentally or deliberately created polyploids.

Model Organisms

Arabidopsis Thaliana

Arabidopsis thaliana, also known as thale cress, has been the model organism for the study of plant genetics. As Drosphila, a species of fruit fly, was to the understanding of early genetics, so has been arabidopsis to the understanding of plant genetics. It was the first plant to ever have its genome sequenced in the year 2000. It has a small genome, making the initial sequencing more attainable. It has a genome size of 125 Mbp that encodes about 25,000 genes. Because an incredible amount of research has been done on the plant, a database called The Arabidopsis Information Resource (TAIR) has been established as a repository for multiple data sets and information on the species. Information housed in TAIR include the complete genome sequence along with gene structure, gene product information, gene expression, DNA and seed stocks, genome maps, genetic and physical markers, publications, and information about the Arabidopsis research community. There are several ecotypes of arabidopsis that have been useful in genetic research, and the natural variation has been used to identify loci important in both biotic and abiotic stress resistance.

Arabidopsis thaliana, growing from between a crack in a sidewalk; it is considered a key model organism in plant genetics.

Brachypodium Distachyon

Brachypodium distachyon is an experimental model grass that has many attributes that make it an excellent model for temperate cereals. Unlike wheat, a tetra or hexaploid species, brachypodium is diploid with a relatively small genome (~355 Mbp) with a short life-cycle, making genomic studies on it simpler.

Nicotiana Benthamiana

Nicotiana benthamiana is often considered a model organism for both plant-pathogen and transgenic studies. Because it is easily transformed with *Agrobacterium tumefaciens*, it is used to study both the expression of pathogen genes introduced into a plant or test new genetic cassette effects.

Genetically Modified Crops

Genetically modified (GM) foods are produced from organisms that have had changes introduced into their DNAusing the methods of genetic engineering. Genetic engineering techniques allow for the introduction of new traits as well as greater control over traits than previous methods such as selective breeding and mutation breeding.

Genetically modifying plants is an important economic activity: in 2017, 89% of corn, 94% of soybeans, and 91% of cotton produced in the US were from genetically modified strains. Since the introduction of GM crops, yields have increased by 22%, and profits have increased to farmers, especially in the developing world, by 68%. An important side effect of GM crops has been decreased land requirements.

Commercial sale of genetically modified foods began in 1994, when Calgene first marketed its unsuccessful Flavr Savr delayed-ripening tomato. Most food modifications have primarily focused on cash crops in high demand by farmers such as soybean, corn, canola, and cotton. Genetically modified crops have been engineered for resistance to pathogens and herbicides and for better nutrient profiles. Other such crops include the economically important GM papaya which are resistant to the highly destructive Papaya ringspot virus, and the nutritionally improved golden rice (it is however still in development).

There is a scientific consensus that currently available food derived from GM crops poses no greater risk to human health than conventional food, but that each GM food needs to be tested on a case-by-case basis before introduction. Nonetheless, members of the public are much less likely than scientists to perceive GM foods as safe. The legal and regulatory status of GM foods varies by country, with some nations banning or restricting them, and others permitting them with widely differing degrees of regulation. There are still ongoing public concerns related to food safety, regulation, labeling, environmental impact, research methods, and the fact that some GM seeds are subject to intellectual propertyrights owned by corporations.

Modern ways to Genetically Modify Plants

Genetic modification has been the cause for much research into modern plant genetics, and has also lead to the sequencing of many plant genomes. Today there are two predominant procedures of transforming genes in organisms: the "Gene gun" method and the *Agrobacterium* method.

Gene Gun Method

The gene gun method is also referred to as "biolistics" (ballistics using biological components). This technique is used for in vivo (within a living organism) transformation and has been especially useful in monocot species like corn and rice. This approach literally shoots genes into plant cells and plant cell chloroplasts. DNA is coated onto small particles of gold or tungsten approximately two micrometres in diameter. The particles are placed in a vacuum chamber and the plant tissue to be engineered is placed below the chamber. The particles are propelled at high velocity using a short pulse of high pressure helium gas, and hit a fine mesh baffle placed above the tissue while the DNA coating continues into any target cell or tissue.

Agrobacterium Method

Transformation via Agrobacterium has been successfully practiced in dicots, i.e. broadleaf plants, such as soybeans and tomatoes, for many years. Recently it has been adapted and is now effective in monocots like grasses, including corn and rice. In general, the *Agrobacterium* method is considered preferable to the gene gun, because of a greater frequency of single-site insertions of the foreign DNA, which allows for easier monitoring. In this method, the tumor inducing (Ti) region is removed from the T-DNA (transfer DNA) and replaced with the desired gene and a marker, which is then inserted into the organism. This may involve direct inoculation of the tissue with a culture of transformed Agrobacterium, or inoculation following treatment with micro-projectile bombardment, which wounds the tissue. Wounding of the target tissue causes the release of phenolic compounds by the plant, which induces invasion of the tissue by Agrobacterium. Because of this, microprojectile bombardment often increases the efficiency of infection with Agrobacterium. The marker is used to find the organism which has successfully taken up the desired gene. Tissues of the organism are then transferred to a medium containing an antibiotic or herbicide, depending on which marker was used. The *Agrobacterium* present is also killed by the antibiotic. Only tissues expressing the marker will survive and possess the gene of interest. Thus, subsequent steps in the process will only use these surviving plants. In order to obtain whole plants from these tissues, they are grown under controlled environmental conditions in tissue culture. This is a process of a series of media, each containing nutrients and hormones. Once the plants are grown and produce seed, the process of evaluating the progeny begins. This process entails selection of the seeds with the desired traits and then retesting and growing to make sure that the entire process has been completed successfully with the desired results.

MICROBIAL GENETICS

Microbial genetics is a branch of genetics concerned with the transmission of hereditary characters in microorganisms. Within the usual definition, microorganisms include prokaryotes like bacteria, unicellular or mycelial eukaryotes e.g., yeasts and other fungi, and viruses, notably bacterial viruses (bacteriophages). Microbial genetics has played a unique role in developing the fields of molecular and cell biology and also has found applications in medicine, agriculture, and the food and pharmaceutical industries.

Because of their relative simplicity, microbes are ideally suited for combined biochemical and genetic studies, and have been successful in providing information on the genetic code and the regulation of gene activity. The operon model formulated by French biologists François Jacob and Jacques Monod in 1961, is one well known example. Based on studies on the induction of enzymes of lactose catabolism in the bacterium Escherichia coli, the operon has provided the groundwork for studies on gene expression and regulation, even up to the present day. The many applications of microbial genetics in medicine and the pharmaceutical industry emerge from the fact that microbes are both the causes of disease and the producers of antibiotics. Genetic studies have been used to understand variation in pathogenic microbes and also to increase the yield of antibiotics from other microbes.

Hereditary processes in microorganisms are analogous to those in multicellular organisms. In both prokaryotic and eukaryotic microbes, the genetic material is DNA; the only known exceptions to this rule are the RNA viruses. Mutations, heritable changes in the DNA, occur spontaneously and the rate of mutation can be increased by mutagenic agents. In practice, the susceptibility of bacteria to mutagenic agents has been used to identify potentially hazardous chemicals in the environment. For example, the Ames test was developed to evaluate the mutagenicity of a chemical in the following way. Plates containing a medium lacking in, for example, the nutrient histidine are inolculated with a histidine requiring strain of the bacterium Salmonella typhimurium. Thus, only cells that revert back to the wild type can grow on the medium. If plates are exposed to a mutagenic agent, the increase in the number of mutants compared with unexposed plates can be observed and a large number of revertants would indicate a strong mutagenic agent. For such studies, microorganisms offer the advantage that they have short mean generation times, are easily cultured in a small space under controlled conditions and have a relatively uncomplicated structure.

Microorganisms, and particularly bacteria, were generally ignored by the early geneticists because of their small in size and apparent lack of easily identifiable variable traits. Therefore, a method of identifying variation and mutation in microbes was fundamental for progress in microbial genetics. As many of the mutations manifest themselves as metabolic abnormalities, methods were developed by which microbial mutants could be detected by selecting or testing for altered phenotypes. Positive selection is defined as the detection of mutant cells and the rejection of unmutated cells. An example of this is the selection of penicillin resistant mutants, achieved by growing organisms in media containing penicillin such that only resistant colonies grow. In contrast, negative selection detects cells that cannot perform a certain function and is used to select mutants that require one or more extra growth factors. Replica plating is used for negative selection and involves two identical prints of colony distributions being made on plates with and without the required nutrients. Those microbes that do not grow on the plate lacking the nutrient can then be selected from the identical plate, which does contain the nutrient.

The first attempts to use microbes for genetic studies were made in the United States shortly before World War II, when George W. Beadle and Edward L. Tatum employed the fungus, Neurospora, to investigate the genetics of tryptophan metabolism and nicotinic acid synthesis. This work led to the development of the "one-gene one-enzyme" hypothesis. Work with bacterial genetics, however, was not really begun until the late 1940s. For a long time, bacteria were thought to lack sexual reproduction, which was believed to be necessary for mixing genes from different individual organisms—a process fundamental for useful genetic studies. However, in 1947, Joshua Lederberg working with Edward Tatum demonstrated the exchange of genetic factors in the bacterium, Escherichia coli. This process of DNA transfer was termed conjugation and requires cell-to-cell contact between two bacteria. It is controlled by genes carried by plasmids, such as the fertility (F) factor, and typically involves the transfer of the plasmid from donor to recipient cell. Other genetic elements, however, including the donor cell chromosome, can sometimes also be mobilized and transferred. Transfer to the host chromosome is rarely complete, but can be used to map the order of genes on a bacterial genome.

Other means by which foreign genes can enter a bacterial cell include transformation, transfection, and transduction. Of the three processes, transformation is probably the most significant. Evidence of transformation in bacteria was first obtained by the British scientist, Fred Griffith in the late 1920s working with Streptococcus pneumoniae and the process was later explained in the 1930s by Oswald Avery and his associates at the Rockefeller Institute in New York. It was discovered that certain bacteria exhibit competence, a state in which cells are able to take up free DNA released by other bacteria. This is the process known as transformation, however, relatively few microorganisms can be naturally transformed.

Certain laboratory procedures were later developed that make it possible to introduce DNA into bacteria, for example electroporation, which modifies the bacterial membrane by treatment with an electric field to facilitate DNA uptake. The latter two processes, transfection and transduction, involve the participation of viruses for nucleic acid transfer. Transfection occurs when bacteria are transformed with DNA extracted from a bacterial virus rather than from another bacterium. Transduction involves the transfer of host genes from one bacterium to another by means of viruses. In generalized transduction, defective virus particles randomly incorporate fragments of the cell DNA; virtually any gene of the donor can be transferred, although the efficiency is low. In specialized transduction, the DNA of a temperate virus excises incorrectly and brings adjacent host genes along with it. Only genes close to the integration point of the virus are transduced, and the efficiency may be high.

After the discovery of DNA transfer in bacteria, bacteria became objects of great interest to geneticists because their rate of reproduction and mutation is higher than in larger organisms; i.e., a mutation occurs in a gene about one time in 10,000,000 gene duplications, and one bacterium may produce 10,000,000,000 offspring in 48 hours. Conjugation, transformation, and transduction have been important methods for mapping the genes on the chromosomes of bacteria. These techniques, coupled with restriction enzyme analysis, cloning DNA sequencing, have allowed for the detailed studies of the bacterial chromosome. Although there are few rules governing gene location, the genes encoding enzymes for many biochemical pathways are often found tightly linked in operons in prokaryotes. Large scale sequencing projects revealed the complete DNA sequence of the genomes of several prokaryotes, even before eukaryotic genomes were considered.

References

- Human-genetics, science: britannica.com, Retrieved 29 March, 2019
- Schwartz M, Vissing J (2002). "Paternal Inheritance of Mitochondrial DNA". N Engl J Med. 347 (8): 576–580. Doi:10.1056/nejmoa020350. PMID 12192017
- Cytogenetics, science: britannica.com, Retrieved 30 April, 2019
- Schwartz, M.; Vissing, J. (2002). "Paternal Inheritance of Mitochondrial DNA". New England Journal of Medicine. 347 (8): 576–80. Doi:10.1056/nejmoa020350. PMID 12192017
- «The Genetic Codes». Www.ncbi.nlm.nih.gov. National Center for Biotechnology Information. Retrieved 16 March 2019

- Behaviour-genetics, science: britannica.com, Retrieved 16 January, 2019
- Leland., Hartwell (2014-09-05). Genetics : from genes to genomes. Goldberg, Michael L., Fischer, Janice A. (Fifth ed.). New York, NY. ISBN 978-0073525310. OCLC 854285781
- Microbial-genetics, anatomy-and-physiology, anatomy-and-physiology, medicine: encyclopedia.com, Retrieved 30 August, 2019
- Kolata, Gina (23 April 2015). "Chinese Scientists Edit Genes of Human Embryos, Raising Concerns". New York Times. Retrieved 24 April 2015
- Ewens W.J. (2004). Mathematical Population Genetics (2nd Edition). Springer-Verlag, New York. ISBN 978-0-387-20191-7

Significant Aspects of Genetics

CHAPTER 3

Some of the fundamental components which are studied in genetics are genomes, nucleic acids, genotypes, phenotypes, alleles, etc. Major concepts that are studied within this field include genetic variation, genome sequencing and gene expression. The topics elaborated in this chapter will help in gaining a better perspective about these components and concepts of genetics.

GENOME

A genome is the complete set of genetic information in an organism. It provides all of the information the organism requires to function. In living organisms, the genome is stored in long molecules of DNA called chromosomes. Small sections of DNA, called genes, code for the RNA and protein molecules required by the organism. In eukaryotes, each cell's genome is contained within a membrane-bound structure called the nucleus. Prokaryotes, which contain no inner membranes, store their genome in a region of the cytoplasm called the nucleoid. The full range of RNA molecules expressed by a genome is known as its transcriptome, and the full assortment of proteins produced by the genome is called its proteome.

There are 23 pairs of chromosomes in the human genome. Between 1990 and 2003, all twenty-three pairs were fully sequenced through an international research undertaking known as the Human Genome Project. The study and analysis of genomes is called genomics.

GENOME SEQUENCING

Genome sequencing is figuring out the order of DNA nucleotides, or bases, in a genome—the order of As, Cs, Gs, and Ts that make up an organism's DNA. The human genome is made up of over 3 billion of these genetic letters.

Today, DNA sequencing on a large scale—the scale necessary for ambitious projects such as sequencing an entire genome—is mostly done by high-tech machines. Much as your eye scans a sequence of letters to read a sentence, these machines "read" a sequence of DNA bases.

A DNA sequence that has been translated from life's chemical alphabet into our alphabet of written letters might look like this:

AGTCCGCGAATACAGGCTCGGT

That is, in this particular piece of DNA, an adenine (A) is followed by a guanine (G), which is followed by a thymine (T), which in turn is followed by a cytosine (C), another cytosine (C), and so on.

By itself, not a whole lot. Genome sequencing is often compared to "decoding," but a sequence is still very much in code. In a sense, a genome sequence is simply a very long string of letters in a mysterious language.

When you read a sentence, the meaning is not just in the sequence of the letters. It is also in the words those letters make and in the grammar of the language. Similarly, the human genome is more than just its sequence.

Imagine the genome as a book written without capitalization or punctuation, without breaks between words, sentences, or paragraphs, and with strings of nonsense letters scattered between and even within sentences.

So sequencing the genome doesn't immediately lay open the genetic secrets of an entire species. Even with a rough draft of the human genome sequence in hand, much work remains to be done. Scientists still have to translate those strings of letters into an understanding of how the genome works: what the various genes that make up the genome do, how different genes are related, and how the various parts of the genome are coordinated. That is, they have to figure out what those letters of the genome sequence mean.

Importance of Genome Sequencing

Sequencing the genome is an important step towards understanding it.

At the very least, the genome sequence will represent a valuable shortcut, helping scientists find genes much more easily and quickly. A genome sequence does contain some clues about where genes are, even though scientists are just learning to interpret these clues.

Scientists also hope that being able to study the entire genome sequence will help them understand how the genome as a whole works—how genes work together to direct the growth, development and maintenance of an entire organism.

Finally, genes account for less than 25 percent of the DNA in the genome, and so knowing the entire genome sequence will help scientists study the parts of the genome outside the genes. This includes the regulatory regions that control how genes are turned on an off, as well as long stretches of "nonsense" or "junk" DNA—so called because we don't yet know what, if anything, it does.

How Do You Sequence a Genome?

The whole genome can't be sequenced all at once because available methods of DNA sequencing can only handle short stretches of DNA at a time.

So instead, scientists must break the genome into small pieces, sequence the pieces, and then reassemble them in the proper order to arrive at the sequence of the whole genome. Much of the work involved in sequencing lies in putting together this giant biological jigsaw puzzle.

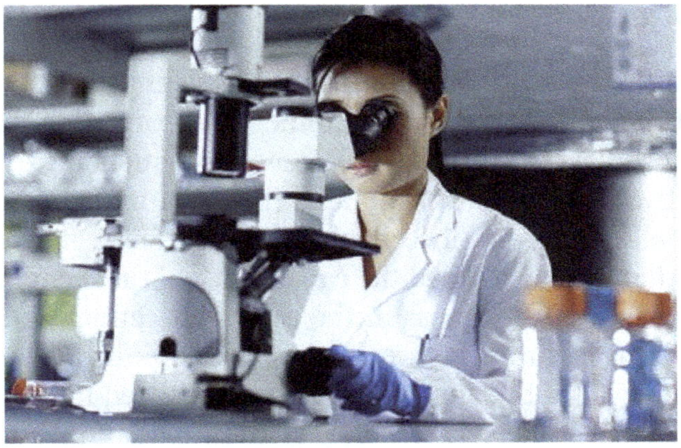

Lab technician working with sequencing machines.

There are two approaches to the task of cutting up the genome and putting it back together again. One strategy, known as the "clone-by-clone" approach, involves first breaking the genome up into relatively large chunks, called clones, about 150,000 base pairs (bp) long. Scientists use genome mapping techniques to figure out where in the genome each clone belongs. Next they cut each clone into smaller, overlapping pieces the right size for sequencing—about 500 BP each. Finally, they sequence the pieces and use the overlaps to reconstruct the sequence of the whole clone.

The other strategy, called "whole-genome shotgun" method, involves breaking the genome up into small pieces, sequencing the pieces, and reassembling the pieces into the full genome sequence.

Room filled with sequencing machines.

Each of these approaches has advantages and disadvantages. The clone-by-clone method is reliable but slow, and the mapping step can be especially time-consuming. By contrast, the whole-genome shotgun method is potentially very fast, but it can be extremely difficult to put together so many tiny pieces of sequence all at once.

Both approaches have already been used to sequence whole genomes. The whole-genome shotgun method was used to sequence the genome of the bacterium Haemophilus influenzae, while the genome of baker's yeast, Saccharomyces cerevisiae, was sequenced with a clone-by-clone method. Sequencing the human genome was done using both approaches.

Whole Genome Sequencing

Whole genome sequencing is the act of deducing the complete nucleic act of deducing the complete nucleic acid sequence of the genetic code, or genome, of an organism or organelle (specifically, the mitochondrion or chloroplast). The first whole genome sequencing efforts, carried out in 1976 and 1977, focused respectively on the bacteriophages (bacteria-infecting viruses) MS2 and ΦX174, which have relatively small genomes. Since then there have been numerous innovations in the field of DNA sequencing that have expanded the capabilities of the technology. Those innovations, combined with increasing cost-effectiveness in the early 21st century, enabled the routine use of whole genome sequencing in laboratories worldwide, which effectively ushered in a new era of biological discovery. The power of the approach has been realized in the study of human populations and human diseases such as cancer, as well as in the elucidation of whole genome sequences of crop plants, livestock, and other species of scientific or agricultural significance. Thus, it is acknowledged generally that there exists great value in a detailed understanding of the nucleic acid sequence—especially the variations in the sequence that correlate with predisposition to health or disease states or with other properties of societal or economic significance in microbial, animal, and plant populations.

Human genome; whole genome sequencing.
Bands of DNA representing a segment of the human genome.

Sequencing Methods: From Genes to Genomes

In 1944 Canadian-born American bacteriologist Oswald Avery and colleagues recognized that the hereditary material passed from parent to offspring was DNA. Subsequent genetic analyses carried out by other scientists on viruses, bacteria, yeast, fruit flies, and nematodes demonstrated that the intentional induction of mutations that disrupted the genetic code, combined with the analysis of observable traits (phenotypes) produced by such mutations, were important approaches to the study of gene function. Such studies, however, were able to query only a fraction of genes in a genome.

The first sequencing methods (the Maxam-Gilbert and Sanger methods), developed in the 1970s, were deployed to reveal the nucleic acid composition of individual genes and the relatively small genomes of certain viruses. The sequencing of larger genomes remained out of reach conceptually, because of high costs and the effort required, until the launch of the Human Genome Project (HGP) in 1990 in the United States. Although the project was not universally embraced, some recognized that technology had evolved to the point where whole genome sequencing of larger genomes could be considered realistically. Particularly important was the development of automated sequencing machines that employed fluorescence instead of radioactive decay for the detection of the sequencing reaction products. Automation offered new possibilities for scaling up the production of DNA sequencing to tackle large genomes.

An early aim of the HGP was to obtain the whole genome sequences of important experimental model organisms, such as the yeast Saccharomyces cerevisiae, the fruit fly Drosophila melanogaster, and the nematode Caenorhabditis elegans. In sequencing those smaller and therefore more-tractable genomes, three outcomes were anticipated. First, the sequences would be of value to the research community, serving to accelerate efforts to understand gene function by using model systems. Second, the experience gained would inform approaches to sequencing the human genome and other similarly sized genomes. Third, functional relationships between sequences of different organisms would be revealed as a consequence of cross-species sequence similarity. Ultimately, with the involvement of more than one thousand scientists worldwide, two human genome sequences were published in 2001. With this development came established methods and analytic standards that were used to sequence other large genomes.

A major challenge for de novo sequencing, in which sequences are assembled for the very first time (such as with the HGP), is the production of individual DNA reads that are of sufficient length and quality to span common repetitive elements, which are a general property of complex genome sequences and a source of ambiguity for sequence assembly. In many of the early de novo whole genome sequencing projects, emphasis was placed on the production of so-called reference sequences, which were of enduring high quality and would serve as the foundation for future experimentation.

An important approach used by many projects that sequenced large genomes involved hierarchical shotgun sequencing, in which segments of genomic DNA were cloned (copied) and arranged into ordered arrays. Those ordered arrays were known as physical maps, and they served to break large genomes into thousands of short DNA fragments. Those short fragments were then aligned, such that identical sequences overlapped, thereby enabling the fragments to be linked together to yield the full-length genomic sequence. The fragments were relatively easy to manipulate in the laboratory, could be apportioned among collaborating laboratories, and were amenable to the detailed error-correction exercises important in generating the high-quality reference sequences sought by HGP scientists. Some genome projects were conducted without the use of such maps, using instead an approach called whole genome shotgun sequencing. This approach avoided the time and expense needed to create physical maps and provided more-rapid access to the DNA sequence.

Whether using physical maps or the whole genome shotgun sequencing approach, the sequencing exercise involved randomly fragmenting either cloned (copied) or native genomic DNA into very short segments that could then be inserted into bacterial cells as plasmids for amplification, producing many copies of the segments, prior to nucleic acid purification and sequence analysis. In a process known as assembly, computer programs were then used to stitch the sequences back together to reconstruct the original DNA sequencing target. Assembly of whole genome shotgun sequencing data was difficult and required sophisticated computer programs and powerful supercomputers, and, even in the years following the completion of the HGP, whole genome shotgun sequence assembly remained a significant challenge for whole genome sequencing projects.

Next-Generation Technologies

Although the first whole genome sequences were in themselves technological and scientific feats of significance, the scientific opportunities and the host of technologies those projects spawned have had even greater impacts. Among the most significant technological developments has been in the area of next-generation DNA sequencing technologies for human genome analysis. Certain of those technologies originally were designed to re-sequence genomes (as opposed to de novo sequencing). In re-sequencing, short sequences are produced and aligned computationally to existing reference genome sequences generated, at least initially, using the older de novo sequencing methods. Next-generation sequencing approaches are characterized generally by the massively parallel production of short sequences, in which multiple DNA fragments are generated simultaneously and in sufficient quantity to redundantly represent every base in the target genome.

ALLELE

An allele is a variant form of a given gene. Sometimes, different alleles can result in different observable phenotypic traits, such as different pigmentation. A notable example of this trait of color variation is Gregor Mendel's discovery that the white and purple flower colors in pea plants were the result of "pure line" traits which could be used as a control for future experiments. However, most genetic variations result in little or no observable variation.

Most multicellular organisms have two sets of chromosomes; that is, they are diploid. In this case, the chromosomes can be paired: each pair is made up of two chromosomes of the same type, known as homologous chromosomes. If both alleles at a gene (or locus) on the homologous chromosomes are the same, they and the organism are homozygous with respect to that gene (or locus). If the alleles are different, they and the organism are heterozygous with respect to that gene.

Alleles that Lead to Dominant or Recessive Phenotypes

In many cases, genotypic interactions between the two alleles at a locus can be described as dominant or recessive, according to which of the two homozygous phenotypes the heterozygote most resembles. Where the heterozygote is indistinguishable from one of the homozygotes, the allele expressed is the one that leads to the "dominant" phenotype, and the other allele is said to be "recessive". The degree and pattern of dominance varies among loci. This type of interaction was first formally described by Gregor Mendel. However, many traits defy this simple categorization and the phenotypes are modeled by co-dominance and polygenic inheritance.

The term "wild type" allele is sometimes used to describe an allele that is thought to contribute to the typical phenotypic character as seen in "wild" populations of organisms, such as fruit flies (*Drosophila melanogaster*). Such a "wild type" allele was historically regarded as leading to a dominant (overpowering - always expressed), common, and normal phenotype, in contrast to "mutant" alleles that lead to recessive, rare, and frequently deleterious phenotypes. It was formerly thought that most individuals were homozygous for the "wild type" allele at most gene loci, and that any alternative "mutant" allele was found in homozygous form in a small minority of "affected" individuals, often as genetic diseases, and more frequently in heterozygous form in "carriers" for the mutant allele. It is now appreciated that most or all gene loci are highly polymorphic, with multiple alleles, whose frequencies vary from population to population, and that a great deal of genetic variation is hidden in the form of alleles that do not produce obvious phenotypic differences.

Multiple Alleles

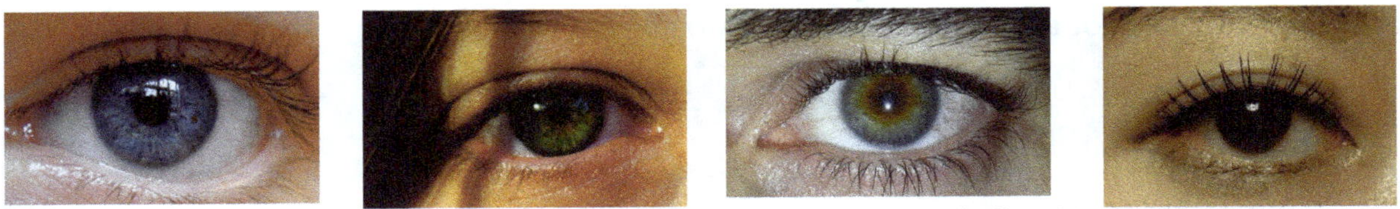

Eye color is an inherited trait influenced by more than one gene, including *OCA2* and *HERC2*. The interaction of multiple genes—and the variation in these genes ("alleles") between individuals—help to determine a person's eye color phenotype.

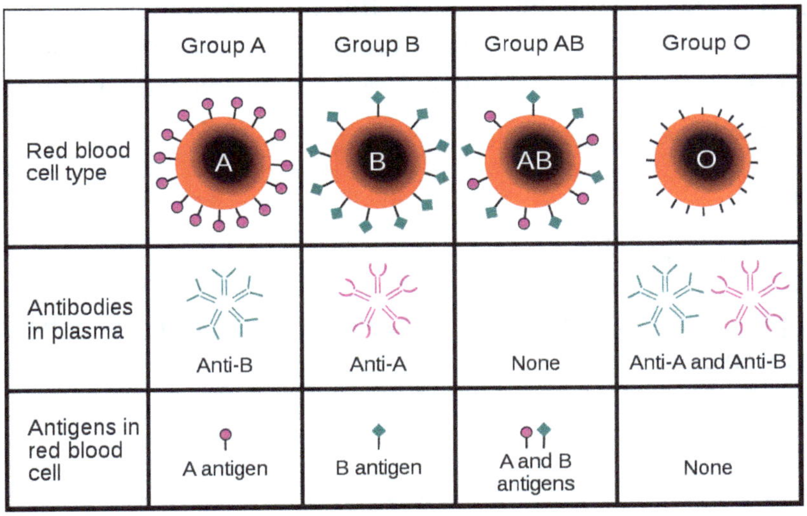

	Group A	Group B	Group AB	Group O
Red blood cell type	A	B	AB	O
Antibodies in plasma	Anti-B	Anti-A	None	Anti-A and Anti-B
Antigens in red blood cell	A antigen	B antigen	A and B antigens	None

In the above figure, ABO blood group system, a person with Type A blood displays A-antigens and may have a genotype I^AI^A or I^Ai. A person with Type B blood displays B-antigens and may have the genotype IBIBor IBi. A person with Type AB blood displays both A- and B-antigens and has the genotype IAIB and a person with Type O blood, displaying neither antigen, has the genotype.

A population or species of organisms typically includes multiple alleles at each locus among various individuals. Allelic variation at a locus is measurable as the number of alleles (polymorphism) present, or the proportion of heterozygotes in the population. A null allele is a gene variant that lacks the gene's normal function because it either is not expressed, or the expressed protein is inactive.

For example, at the gene locus for the ABO blood typecarbohydrate antigens in humans, classical genetics recognizes three alleles, I^A, I^B, and i, which determine compatibility of blood transfusions. Any individual has one of six possible genotypes (I^AI^A, I^Ai, I^BI^B, I^Bi, I^AI^B, and ii) which produce one of four possible phenotypes: "Type A" (produced by I^AI^A homozygous and I^Ai heterozygous genotypes), "Type B" (produced by I^BI^B homozygous and I^Bi heterozygous genotypes), "Type AB" produced by I^AI^Bheterozygous genotype, and "Type O" produced by ii homozygous genotype. (It is now known that each of the A, B, and O alleles is actually a class of multiple alleles with different DNA sequences that produce proteins with identical properties: more than 70 alleles are known at the ABO locus. Hence an individual with "Type A" blood may be an AO heterozygote, an AA homozygote, or an AA heterozygote with two different "A" alleles).

Genotype Frequencies

The frequency of alleles in a diploid population can be used to predict the frequencies of the corresponding genotypes. For a simple model, with two alleles:

$$p + q = 1$$

$$p^2 + 2pq + q^2 = 1$$

where *p* is the frequency of one allele and *q* is the frequency of the alternative allele, which necessarily sum to unity. Then, p^2 is the fraction of the population homozygous for the first allele, $2pq$ is the fraction of heterozygotes, and q^2 is the fraction homozygous for the alternative allele. If the first allele is dominant to the second then the fraction of the population that will show the dominant phenotype is $p^2 + 2pq$, and the fraction with the recessive phenotype is q^2.

With three alleles:

$$p + q + r = 1$$

$$p^2 + q^2 + r^2 + 2pq + 2pr + 2qr = 1.$$

In the case of multiple alleles at a diploid locus, the number of possible genotypes (G) with a number of alleles (a) is given by the expression:

$$G = \frac{a(a+1)}{2}$$

Allelic Dominance in Genetic Disorders

A number of genetic disorders are caused when an individual inherits two recessive alleles for a single-gene trait. Recessive genetic disorders include albinism, cystic fibrosis, galactosemia, phenylketonuria (PKU), and Tay–Sachs disease. Other disorders are also due to recessive alleles, but because the gene locus is located on the X chromosome, so that males have only one copy (that is, they are hemizygous), they are more frequent in males than in females. Examples include red-green color blindness and fragile X syndrome.

Other disorders, such as Huntington's disease, occur when an individual inherits only one dominant allele.

Epialleles

While heritable traits are typically studied in terms of genetic alleles, epigenetic marks such as DNA methylationcan be inherited at specific genomic regions in certain species, a process termed transgenerational epigenetic inheritance. The term *epiallele* is used to distinguish these heritable marks from traditional alleles, which are defined by nucleotide sequence. A specific class of epiallele, the metastable epialleles, has been discovered in mice and in humans which is characterized by stochastic (probabilistic) establishment of epigenetic state that can be mitotically inherited.

Dominance (Genetics)

Dominance, in genetics, is the phenomenon of one variant (allele) of a gene on a chromosome masking or overriding the effect of a different variant of the same gene on the other copy of the chromosome. The first variant is termed dominant and the second recessive. This state of having two different variants of the same gene on each chromosome is originally caused by a mutationin one of the genes, either new (*de novo*) or inherited. The terms autosomal dominant or autosomal recessive are used to describe gene variants on non-sex chromosomes (autosomes) and their associated traits, while those on sex chromosomes (allosomes) are termed X-linked dominant, X-linked recessive or Y-linked, and these show a very different inheritance and presentation pattern to autosomal traits which depends on the sex of the individual. Additionally, there are other forms of dominance such as incomplete dominance, in which a gene variant has a partial effect compared to when it is present on both chromosomes, and co-dominance, in which different variants on each chromosome both show their associated traits.

Inheritance of dwarfing in maize. Demonstrating the heights of plants from the two parent variations and their F1 heterozygous hybrid (centre).

Dominance is not inherent to an allele or its traits (phenotype). It is a strictly relative effect between two alleles of a given gene of any function; one allele can be dominant over a second allele of the same gene, recessive to a third and co-dominant with a fourth. Additionally, one allele may be dominant for one trait but not others.

Often, the dominant allele codes for a functional protein whereas the recessive allele does not. Hence, many autosomal recessive conditions are the result of lack of a functioning protein, for example an enzyme. On the other hand, many autosomal dominant conditions result in the presence of a diseased form of a critical protein or simply not enough expressivity of the critical protein.

Dominance is a key concept in Mendelian inheritance and classical genetics. Letters and Punnett squares are used to demonstrate the principles of dominance in teaching, and the use of upper case letters for dominant alleles and lower case letters for recessive alleles is a widely followed convention. A classic example of dominance is the inheritance of seed shape in peas. Peas may be round, associated with allele *R*, or wrinkled, associated with allele *r*. In this case, three combinations of alleles (genotypes) are possible: *RR*, *Rr*, and *rr*. The *RR* (homozygous) individuals have round peas, and the *rr* (homozygous) individuals have wrinkled peas. In *Rr*(heterozygous) individuals, the *R* allele masks the presence of the *r* allele, so these individuals also have round peas. Thus, allele *R* is dominant over allele *r*, and allele *r* is recessive to allele *R*.

Dominance differs from epistasis, the phenomenon of an allele of one gene masking the effect of alleles of a *different* gene.

The concept of dominance was introduced by Gregor Johann Mendel. Though Mendel, "The Father of Genetics", first used the term in the 1860s, it was not widely known until the early twentieth century. Mendel observed that, for a variety of traits of garden peas having to do with the appearance of seeds, seed pods, and plants, there were two discrete phenotypes, such as round versus wrinkled seeds, yellow versus green seeds, red versus white flowers or tall versus short plants. When bred separately, the plants always produced the same phenotypes, generation after generation. However, when lines with different phenotypes were crossed (interbred), one and only one of the parental phenotypes showed up in the offspring (green, or round, or red, or tall). However, when these hybrid plants were crossed, the offspring plants showed the two original phenotypes, in a characteristic 3:1 ratio, the more common phenotype being that of the parental hybrid plants. Mendel reasoned that each parent in the first cross was a homozygote for different alleles (one parent AA and the other parent aa), that each contributed one allele to the offspring, with the result that all of these hybrids were heterozygotes (Aa), and that one of the two alleles in the hybrid cross dominated expression of the other: A masked a. The final cross between two heterozygotes (Aa X Aa) would produce AA, Aa, and aa offspring in a 1:2:1 genotype ratio with the first two classes showing the (A) phenotype, and the last showing the (a) phenotype, thereby producing the 3:1 phenotype ratio.

Mendel did not use the terms gene, allele, phenotype, genotype, homozygote, and heterozygote, all of which were introduced later. He did introduce the notation of capital and lowercase letters for dominant and recessive alleles, respectively, still in use today.

Chromosomes, Genes and Alleles

Most animals and some plants have paired chromosomes, and are described as diploid. They have two versions of each chromosome, one contributed by the mother's ovum, and the other by the father's sperm, known as gametes, described as haploid, and created through meiosis. These gametes then fuse during fertilization during sexual reproduction, into a new single cell zygote, which divides multiple times, resulting in a new organism with the same number of pairs of chromosomes in each (non-gamete) cell as its parents.

Each chromosome of a matching (homologous) pair is structurally similar to the other, and has a very similar DNA sequence (loci, singular locus). The DNA in each chromosome functions as a series of discrete genes that influence various traits. Thus, each gene also has a corresponding homologue, which may exist in different versions called alleles. The alleles at the same locus on the two homologous chromosomes may be identical or different.

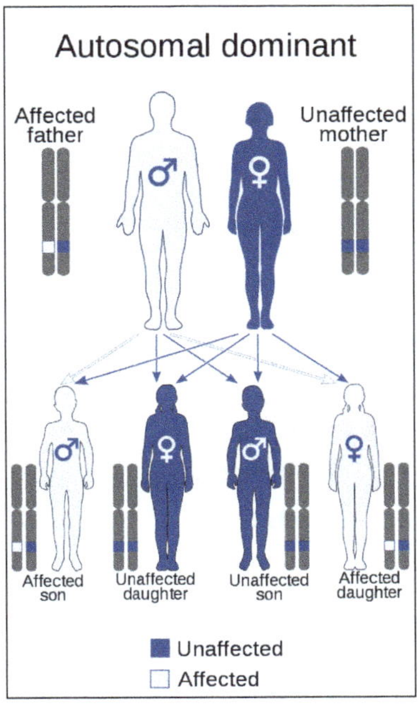

An autosomal dominant pattern.

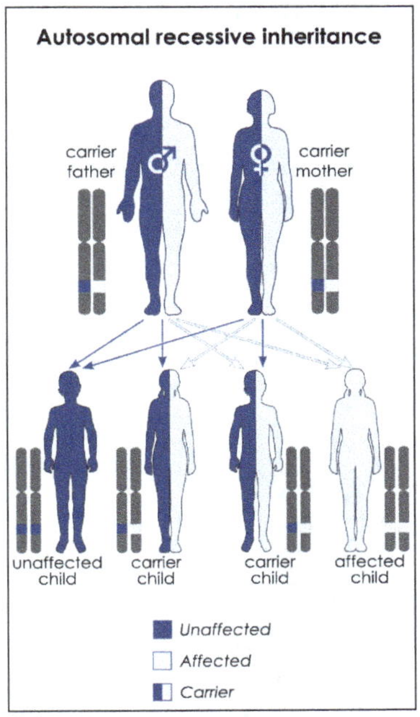

An autosomal recessive pattern.

The blood type of a human is determined by a gene that creates an A, B, AB or O blood type and is located in the long arm of chromosome nine. There are three different alleles that could be present at this locus, but only two can be present in any individual, one inherited from their mother and one from their father.

If two alleles of a given gene are identical, the organism is called a homozygote and is said to be homozygous with respect to that gene; if instead the two alleles are different, the organism is a heterozygote and is heterozygous. The genetic makeup of an organism, either at a single locus or over all its genes collectively, is called its genotype. The genotype of an organism directly and indirectly affects its molecular, physical, and other traits, which individually or collectively are called its phenotype. At heterozygous gene loci, the two alleles interact to produce the phenotype.

Dominance

Complete Dominance

In complete dominance, the effect of one allele in a heterozygous genotype completely masks the effect of the other. The allele that masks the other is said to be *dominant* to the latter, and the allele that is masked is said to be *recessive* to the former. Complete dominance, therefore, means that the phenotype of the heterozygote is indistinguishable from that of the dominant homozygote.

A classic example of dominance is the inheritance of seed shape (pea shape) in peas. Peas may be round (associated with allele *R*) or wrinkled (associated with allele *r*). In this case, three combinations of alleles (genotypes) are possible: *RR* and *rr* are homozygous and *Rr* is heterozygous. The *RR* individuals have round peas and the *rr* individuals have wrinkled peas. In *Rr* individuals the *R* allele masks the presence of the *r* allele, so these individuals also have round peas. Thus, allele *R* is completely dominant to allele *r*, and allele *r* is recessive to allele *R*.

Incomplete Dominance

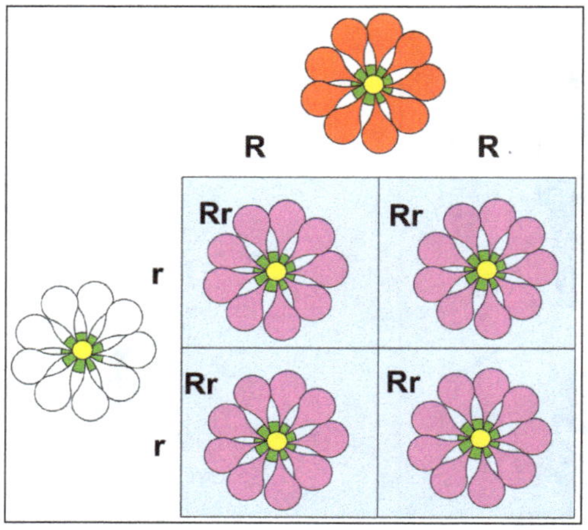

This Punnett square illustrates incomplete dominance. In this example, the red petal trait associated with the R allele recombines with the white petal trait of the r allele. The plant incompletely expresses the dominant trait (R) causing plants with the Rr genotype to express flowers with less red pigment resulting in pink flowers. The colors are not blended together, the dominant trait is just expressed less strongly.

Incomplete dominance (also called *partial dominance, semi-dominance*or *intermediate inheritance*) occurs when the phenotype of the heterozygous genotype is distinct from and often intermediate to the phenotypes of the homozygous genotypes. For example, the snapdragon flower color is homozygous for either red or white. When the red homozygous flower is paired with the white homozygous flower, the result yields a pink snapdragon flower. The pink snapdragon is the result of incomplete dominance. A similar type of incomplete dominance is found in the four o'clock plant wherein pink color is produced when true-bred parents of white and red flowers are crossed. In quantitative genetics, where phenotypes are measured and treated numerically, if a heterozygote's phenotype is exactly between (numerically) that of the two homozygotes, the phenotype is said to exhibit *no dominance* at all, i.e. dominance exists only when the heterozygote's phenotype measure lies closer to one homozygote than the other.

When plants of the F_1 generation are self-pollinated, the phenotypic and genotypic ratio of the F_2 generation will be 1:2:1 (Red:Pink:White).

Co-dominance

Co-dominance occurs when the contributions of both alleles are visible in the phenotype.

For example, in the ABO blood group system, chemical modifications to a glycoprotein(the H antigen) on the surfaces of blood cells are controlled by three alleles, two of which are co-dominant to each other (I^A, I^B) and dominant over the recessive *i* at the ABO locus. The I^A and I^B alleles produce different modifications. The enzyme coded for by I^A adds an N-acetylgalactosamine to the membrane-bound H antigen. The I^B enzyme adds a galactose. The *i* allele produces no modification.

Thus I^A and I^B alleles are each dominant to i (I^AI^A and I^Aiindividuals both have type A blood, and I^BI^B and I^Bi individuals both have type B blood, but I^AI^B individuals have both modifications on their blood cells and thus have type AB blood, so the I^A and I^B alleles are said to be co-dominant).

Co-dominance in a Camellia cultivar.

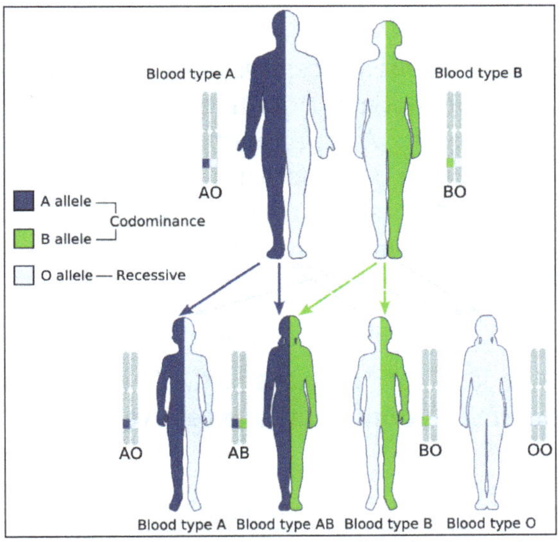

A and B blood types in humans show co-dominance, but the O type is recessive to A and B.

Another example occurs at the locus for the beta-globin component of hemoglobin, where the three molecular phenotypes of Hb^A/Hb^A, Hb^A/Hb^S, and Hb^S/Hb^S are all distinguishable by protein electrophoresis. (The medical condition produced by the heterozygous genotype is called *sickle-cell trait* and is a milder condition distinguishable from *sickle-cell anemia*, thus the alleles show *incomplete dominance* with respect to anemia, see above). For most gene loci at the molecular level, both alleles are expressed co-dominantly, because both are transcribed into RNA.

Co-dominance, where allelic products co-exist in the phenotype, is different from incomplete dominance, where the quantitative interaction of allele products produces an intermediate phenotype. For example, in co-dominance, a red homozygous flower and a white homozygous flower will produce offspring that have red and white spots. When plants of the F1 generation are self-pollinated, the phenotypic and genotypic ratio of the F2 generation will be 1:2:1 (Red:Spotted:White). These ratios are the same as those for incomplete dominance. Again, note that this classical terminology is inappropriate – in reality such cases should not be said to exhibit dominance at all.

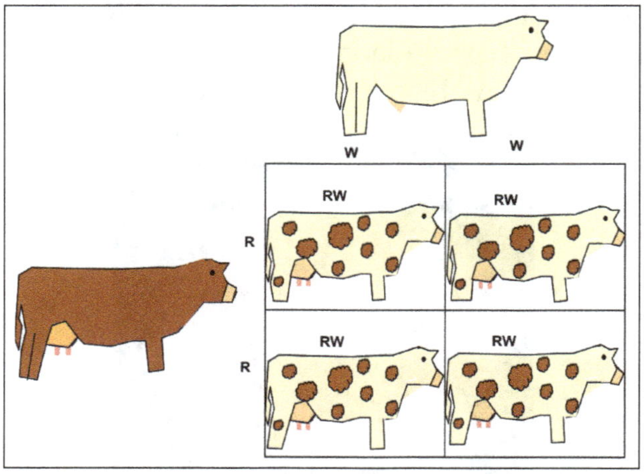

This diagram shows co-dominance. In this example a white bull (WW) mates with a red cow (RR), and their offspring exhibit co-dominance expressing both white and red hairs.

Addressing Common Misconceptions

While it is often convenient to talk about a *recessive allele* or a *dominant trait*, dominance is not inherent to either an allele or its phenotype. Dominance is a relationship between two alleles of a gene and their associated phenotypes. A "dominant" allele is dominant to a particular allele of the same gene that can be inferred from the context, but it may be recessive to a third allele, and codominant to a fourth. Similarly, a "recessive" trait is a trait associated with a particular recessive allele implied by the context, but that same trait may occur in a different context where it is due to some other gene and a dominant allele.

Dominance is unrelated to the nature of the phenotype itself, that is, whether it is regarded as "normal" or "abnormal," "standard" or "nonstandard," "healthy" or "diseased," "stronger" or "weaker," or more or less extreme. A dominant or recessive allele may account for any of these trait types.

Dominance does not determine whether an allele is deleterious, neutral or advantageous. However, selectionmust operate on genes indirectly through phenotypes, and dominance affects the exposure of alleles in phenotypes, and hence the rate of change in allele frequencies under selection. Deleterious recessive alleles may persist in a population at low frequencies, with most copies carried in heterozygotes, at no cost to those individuals. These rare recessives are the basis for many hereditary genetic disorders.

Dominance is also unrelated to the distribution of alleles in the population. Both dominant and recessive alleles can be extremely common or extremely rare.

Nomenclature

In genetics, symbols began as algebraic placeholders. When one allele is dominant to another, the oldest convention is to symbolize the dominant allele with a capital letter. The recessive allele is assigned the same letter in lower case. In the pea example, once the dominance relationship between the two alleles is known, it is possible to designate the dominant allele that produces a round shape by a capital-letter symbol R, and the recessive allele that produces a wrinkled shape by a lower-case symbol r. The homozygous dominant, heterozygous, and homozygous recessive genotypes are then written RR, Rr, and rr, respectively. It would also be possible to designate the two alleles as W and w, and the three genotypes WW, Ww, and ww, the first two of which produced round peas and the third wrinkled peas. Note that the choice of "R" or "W" as the symbol for the dominant allele does not pre-judge whether the allele causing the "round" or "wrinkled" phenotype when homozygous is the dominant one.

A gene may have several alleles. Each allele is symbolized by the locus symbol followed by a unique superscript. In many species, the most common allele in the wild population is designated the wild type allele. It is symbolized with a + character as a superscript. Other alleles are dominant or recessive to the wild type allele. For recessive alleles, the locus symbol is in lower case letters. For alleles with any degree of dominance to the wild type allele, the first letter of the locus symbol is in upper case. For example, here are some of the alleles at the *a* locus of the laboratory mouse, *Mus musculus*: A^y, dominant yellow; a^+, wild type; and a^{bt}, black and tan. The a^{bt} allele is recessive to the wild type allele, and the A^y allele is codominant to the wild type allele. The A^yallele is also codominant to the a^{bt} allele, but showing that relationship is beyond the limits of the rules for mouse genetic nomenclature.

Rules of genetic nomenclature have evolved as genetics has become more complex. Committees have standardized the rules for some species, but not for all. Rules for one species may differ somewhat from the rules for a different species.

Relationship to other Genetic Concepts

Multiple Alleles

Although any individual of a diploid organism has at most two different alleles at any one locus (barring aneuploidies), most genes exist in a large number of allelic versions in the population as a whole. If the alleles have different effects on the phenotype, sometimes their dominance relationships can be described as a series.

For example, coat color in domestic cats is affected by a series of alleles of the *TYR* gene (which encodes the enzyme tyrosinase). The alleles C, c^b, c^s, and c^a (full colour, Burmese, Siamese, and albino, respectively) produce different levels of pigment and hence different levels of colour dilution. The C allele (full colour) is completely dominant over the last three and the c^a allele (albino) is completely recessive to the first three.

Autosomal vs. Sex-linked Dominance

In humans and other mammal species, sex is determined by two sex chromosomes called the X chromosomeand the Y chromosome. Human females are typically XX; males are typically XY. The remaining pairs of chromosome are found in both sexes and are called autosomes; genetic traits due to loci on these chromosomes are described as autosomal, and may be dominant or recessive. Genetic traits on the X and Ychromosomes are called sex-linked, because they are linked to sex chromosomes, not because they are characteristic of one sex or the other. In practice, the term almost always refers to X-linked traits and a great many such traits (such as red-green colour vision deficiency) are not affected by sex. Females have two copies of every gene locus found on the

X chromosome, just as for the autosomes, and the same dominance relationships apply. Males, however, have only one copy of each X chromosome gene locus, and are described as hemizygous for these genes. The Y chromosome is much smaller than the X, and contains a much smaller set of genes, including, but not limited to, those that influence 'maleness', such as the SRY gene for testis determining factor. Dominance rules for sex-linked gene loci are determined by their behavior in the female: because the male has only one allele (except in the case of certain types of Y chromosome aneuploidy), that allele is always expressed regardless of whether it is dominant or recessive. Birds have oppositely sex chromosomes: male birds have ZZ and female birds ZW chromosomes. However, inheritance of traits reminds XY-system otherwise; male zebra finches may carry white colouring gene in their one of two Z chromosome, but females develop white colouring always. Grasshoppers have XO-system. Females have XX, but males only X. There is no Y chromosome at all.

Epistasis

Epistasis ["*epi* + *stasis* = to sit on top"] is an interaction between alleles at two *different* gene loci that affect a single trait, which may sometimes resemble a dominance interaction between two *different* alleles at the *same*locus. Epistasis modifies the characteristic 9:3:3:1 ratio expected for two non-epistatic genes. For two loci, 14 classes of epistatic interactions are recognized. As an example of *recessive epistasis*, one gene locus may determine whether a flower pigment is yellow (AA or Aa) or green (aa), while another locus determines whether the pigment is produced (BB or Bb) or not (bb). In a bb plant, the flowers will be white, irrespective of the genotype of the other locus as AA, Aa, or aa. The bb combination is *not* dominant to the A allele: rather, the Bgene shows *recessive epistasis* to the A gene, because the B locus when homozygous for the *recessive* allele (bb) suppresses phenotypic expression of the A locus. In a cross between two AaBb plants, this produces a characteristic 9:3:4 ratio, in this case of yellow : green : white flowers.

In *dominant epistasis*, one gene locus may determine yellow or green pigment as in the previous example: AAand Aa are yellow, and aa are green. A second locus determines whether a pigment precursor is produced (dd) or not (DD or Dd). Here, in a DD or Dd plant, the flowers will be colorless irrespective of the genotype at the *A*locus, because of the epistatic effect of the dominant D allele. Thus, in a cross between two AaDd plants, 3/4 of the plants will be colorless, and the yellow and green phenotypes are expressed only in dd plants. This produces a characteristic 12:3:1 ratio of white : yellow : green plants.

Supplementary epistasis occurs when two loci affect the same phenotype. For example, if pigment color is produced by CC or Cc but not cc, and by DD or Dd but not dd, then pigment is not produced in any genotypic combination with either cc *or* dd. That is, *both* loci must have at least one dominant allele to produce the phenotype. This produces a characteristic 9:7 ratio of pigmented to unpigmented plants. *Complementary epistasis* in contrast produces an unpigmented plant if and only if the genotype is cc *and* dd, and the characteristic ratio is 15:1 between pigmented and unpigmented plants.

Classical genetics considered epistatic interactions between two genes at a time. It is now evident from molecular genetics that all gene loci are involved in complex interactions with many other genes (e.g., metabolic pathways may involve scores of genes), and that this creates epistatic interactions that are much more complex than the classic two-locus models.

Hardy–Weinberg Principle

The frequency of the heterozygous state (which is the carrier state for a recessive trait) can be estimated using the Hardy–Weinberg formula:

$$p^2 + 2pq + q^2 = 1$$

This formula applies to a gene with exactly two alleles and relates the frequencies of those alleles in a large population to the frequencies of their three genotypes in that population.

For example, if p is the frequency of allele A, and q is the frequency of allele a then the terms p^2, $2pq$, and $q^2$are the frequencies of the genotypes AA, Aa and aa respectively. Since the gene has only two alleles, all alleles must be either A or a and $p + q = 1$. Now, if A is completely dominant to a then the frequency of the carrier genotype Aa cannot be directly observed (since it has the same traits as the homozygous genotype AA), however it can be estimated from the frequency of the recessive trait in the population, since this is the same as that of the homozygous genotype aa. i.e. the individual allele frequencies can be estimated: $q = \sqrt{f(\text{aa})}$, $p = 1 - q$, and from those the frequency of the carrier genotype can be derived: $f(\text{Aa}) = 2pq$.

This formula relies on a number of assumptions and an accurate estimate of the frequency of the recessive trait. In general, any real-world situation will deviate from these assumptions to some degree, introducing corresponding inaccuracies into the estimate. If the recessive trait is rare, then it will be hard to estimate its frequency accurately, as a very large sample size will be needed.

Dominant versus Advantageous

The property of "dominant" is sometimes confused with the concept of advantageous and the property of "recessive" is sometimes confused with the concept of deleterious, but the phenomena are distinct. Dominance describes the phenotype of heterozygotes with regard to the phenotypes of the homozygotes and without respect to the degree to which different phenotypes may be beneficial or deleterious. Since many genetic disease alleles are recessive and because the word dominance has a positive connotation, the assumption that the dominant phenotype is superior with respect to fitness is often made. This is not assured however; as discussed below while most genetic disease alleles are deleterious and recessive, not all genetic diseases are recessive.

Nevertheless, this confusion has been pervasive throughout the history of genetics and persists to this day. Addressing this confusion was one of the prime motivations for the publication of the Hardy-Weinberg principle.

Molecular Mechanisms

The molecular basis of dominance was unknown to Mendel. It is now understood that a gene locus includes a long series (hundreds to thousands) of bases or nucleotides of deoxyribonucleic acid (DNA) at a particular point on a chromosome. The central dogma of molecular biology states that "*DNA makes RNA makes protein*", that is, that DNA is transcribed to make an RNA copy, and RNA is translated to make a protein. In this process, different alleles at a locus may or may not be transcribed, and if transcribed may be translated to slightly different versions of the same protein (called isoforms). Proteins often function as enzymes that catalyze chemical reactions in the cell, which directly or indirectly produce phenotypes. In any diploid organism, the DNA sequences of the two alleles present at any gene locus may be identical (homozygous) or different (heterozygous). Even if the gene locus is heterozygous at the level of the DNA sequence, the proteins made by each allele may be identical. In the absence of any difference between the protein products, neither allele can be said to be dominant. Even if the two protein products are slightly different (allozymes), it is likely that they produce the same phenotype with respect to enzyme action, and again neither allele can be said to be dominant.

Loss of Function and Haplosufficiency

Dominance typically occurs when one of the two alleles is non-functional at the molecular level, that is, it is not transcribed or else does not produce a functional protein product. This can be the result of a mutation that alters the DNA sequence of the allele. An organism homozygous for the non-functional allele will generally show a distinctive phenotype, due to the absence of the protein product. For example, in humans and other organisms, the unpigmented skin of the albino phenotype results when an individual is homozygous for an allele that encodes a non-functional version of an enzyme needed to produce the skin pigment melanin. It is important to understand that it is not the lack of function that allows the allele to be described as recessive: this is the interaction with the alternative allele in the heterozygote. Three general types of interaction are possible:

1. In the typical case, the single functional allele makes sufficient protein to produce a phenotype identical to that of the homozygote: this is called *haplosufficiency*. For example, suppose the standard amount of enzyme produced in the functional homozygote is 100%, with the two functional alleles contributing 50% each. The single functional allele in the heterozygote produces 50% of the standard amount of enzyme, which is sufficient to produce the standard phenotype. If the heterozygote and the functional-allele homozygote have identical phenotypes, the functional allele is dominant to the non-functional allele. This occurs at the albino gene locus: the heterozygote produces sufficient enzyme to convert the pigment precursor to melanin, and the individual has standard pigmentation.

2. Less commonly, the presence of a single functional allele gives a phenotype that is not normal but less severe than that of the non-functional homozygote. This occurs when the functional allele is not haplo-sufficient. The terms haplo-insufficiency and incomplete dominance are typically applied to these cases. The intermediate interaction occurs where the heterozygous genotype produces a phenotype intermediate between the two homozygotes. Depending on which of the two homozygotes the heterozygote most resembles, one allele is said to show *incomplete dominance* over the other. For example, in humans the *Hb* gene locus is responsible for the Beta-chain protein (HBB) that is one of the two globin proteins that make up the blood pigment hemoglobin. Many people are homozygous for an allele called Hb^A; some persons carry an alternative allele called Hb^S, either as homozygotes or heterozygotes. The hemoglobin molecules of Hb^S/Hb^S homozygotes undergo a change in shape that distorts the morphology of the red blood cells, and causes a severe, life-threatening form of anemiacalled sickle-cell anemia. Persons heterozygous Hb^A/Hb^S for this allele have a much less severe form of anemia called

sickle-cell trait. Because the disease phenotype of Hb^A/Hb^S heterozygotes is more similar to but not identical to the Hb^A/Hb^A homozygote, the Hb^A allele is said to be *incompletely dominant* to the Hb^S allele.

3. Rarely, a single functional allele in the heterozygote may produce insufficient gene product for any function of the gene, and the phenotype resembles that of the homozygote for the non-functional allele. This complete *haploinsufficiency* is very unusual. In these cases, the non-functional allele would be said to be dominant to the functional allele. This situation may occur when the non-functional allele produces a defective protein that interferes with the proper function of the protein produced by the standard allele. The presence of the defective protein "dominates" the standard protein, and the disease phenotype of the heterozygote more closely resembles that of the homozygote for two defective alleles. Note that the term dominant is often incorrectly applied to defective alleles whose homozygous phenotype has not been examined, but which cause a distinct phenotype when heterozygous with the normal allele. This phenomenon occurs in a number of trinucleotide repeat diseases, one example being Huntington's disease.

Dominant-negative Mutations

Many proteins are normally active in the form of a multimer, an aggregate of multiple copies of the same protein, otherwise known as a homomultimeric protein or homooligomeric protein. In fact, a majority of the 83,000 different enzymes from 9800 different organisms in the BRENDA Enzyme Database represent homooligomers. When the wild-type version of the protein is present along with a mutant version, a mixed multimer can be formed. A mutation that leads to a mutant protein that disrupts the activity of the wild-type protein in the multimer is a dominant-negative mutation.

A dominant-negative mutation may arise in a human somatic cell and provide a proliferative advantage to the mutant cell, leading to its clonal expansion. For instance, a dominant-negative mutation in a gene necessary for the normal process of programmed cell death (Apoptosis) in response to DNA damage can make the cell resistant to apoptosis. This will allow proliferation of the clone even when excessive DNA damage is present. Such dominant-negative mutations occur in the tumor suppressor gene *p53*. The P53 wild-type protein is normally present as a four-protein multimer (oligotetramer). Dominant-negative *p53* mutations occur in a number of different types of cancer and pre-cancerous lesions (e.g. brain tumors, breast cancer, oral pre-cancerous lesions and oral cancer).

Dominant-negative mutations also occur in other tumor suppressor genes. For instance two dominant-negative germ line mutations were identified in the Ataxia telangiectasia mutated (ATM) gene which increases susceptibility to breast cancer. Dominant negative mutations of the transcription factor C/EBPα can cause acute myeloid leukemia. Inherited dominant negative mutations can also increase the risk of diseases other than cancer. Dominant-negative mutations in Peroxisome proliferator-activated receptor gamma (PPARγ) are associated with severe insulin resistance, diabetes mellitus and hypertension.

Dominant-negative mutations have also been described in organisms other than humans. In fact, the first study reporting a mutant protein inhibiting the normal function of a wild-type protein in a mixed multimer was with the bacteriophage T4 tail fiber protein GP37. Mutations that produce a truncated protein rather than a full-length mutant protein seem to have the strongest dominant-negative effect in the studies of P53, ATM, C/EBPα, and bacteriophage T4 GP37.

Dominant and Recessive Genetic Diseases in Humans

In humans, many genetic traits or diseases are classified simply as "dominant" or "recessive". Especially with so-called recessive diseases, which are indeed a factor of recessive genes, but can oversimplify the underlying molecular basis and lead to misunderstanding of the nature of dominance. For example, the recessive genetic disease phenylketonuria (PKU) results from any of a large number (>60) of alleles at the gene locus for the enzyme phenylalanine hydroxylase (PAH). Many of these alleles produce little or no PAH, as a result of which the substrate phenylalanine (Phe) and its metabolic byproducts accumulate in the central nervous system and can cause severe intellectual disability if untreated.

The genotypes and phenotypic consequences of interactions among three alleles are shown in the following table:

Genotype	PAH activity	[Phe] conc	PKU ?
AA	100%	60 uM	No
AB	30%	120 uM	No
CC	5%	200 ~ 300 uM	Hyperphenylalaninemia
BB	0.3%	600 ~ 2400 uM	Yes

In unaffected persons homozygous for a standard functional allele (AA), PAH activity is standard (100%), and the concentration of phenylalanine in the blood [Phe] is about 60 uM. In untreated persons homozygous for one of the PKU alleles (BB), PAH activity is close to zero, [Phe] ten to forty times standard, and the individual manifests PKU.

In the AB heterozygote, PAH activity is only 30% (not 50%) of standard, blood [Phe] is elevated two-fold, and the person does not manifest PKU. Thus, the A allele is dominant to the B allele with respect to PKU, but the Ballele is incompletely dominant to the A allele with respect to its molecular effect, determination of PAH activity level (0.3% < 30% << 100%). Finally, the A allele is an incomplete dominant to B with respect to [Phe], as 60 uM < 120 uM << 600 uM. Note once more that it is irrelevant to the question of dominance that the recessive allele produces a more extreme [Phe] phenotype.

For a third allele C, a CC homozygote produces a very small amount of PAH enzyme, which results in a somewhat elevated level of [Phe] in the blood, a condition called hyperphenylalaninemia, which does not result in intellectual disability.

That is, the dominance relationships of any two alleles may vary according to which aspect of the phenotype is under consideration. It is typically more useful to talk about the phenotypic consequences of the allelic interactions involved in any genotype, rather than to try to force them into dominant and recessive categories.

NUCLEIC ACID

Nucleic acid is a naturally occurring chemical compound that is capable of being broken down to yield phosphoric acid, sugars, and a mixture of organic bases (purines and pyrimidines). Nucleic acids are the main information-carrying molecules of the cell, and, by directing the process of protein synthesis, they determine the inherited characteristics of every living thing. The two main classes of nucleic acids are deoxyribonucleic acid (DNA) and ribonucleic acid (RNA). DNA is the master blueprint for life and constitutes the genetic material in all free-living organisms and most viruses. RNA is the genetic material of certain viruses, but it is also found in all living cells, where it plays an important role in certain processes such as the making of proteins.

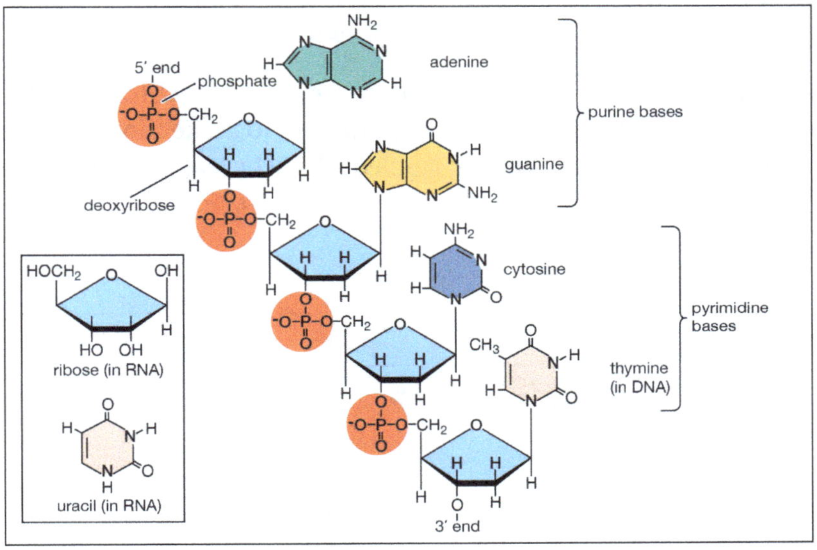

Portion of polynucleotide chain of deoxyribonucleic acid (DNA). The inset shows the corresponding pentose sugar and pyrimidine base in ribonucleic acid (RNA).

Nucleotides: Building Blocks of Nucleic Acids

Basic Structure

Nucleic acids are polynucleotides—that is, long chainlike molecules composed of a series of nearly identical building blocks called nucleotides. Each nucleotide consists of a nitrogen-containing aromatic base attached to a pentose (five-carbon) sugar, which is in turn attached to a phosphate group. Each nucleic acid contains four of five possible nitrogen-containing bases: adenine (A), guanine (G), cytosine (C), thymine (T), and uracil (U). A and G are categorized as purines, and C, T, and U are collectively called pyrimidines. All nucleic acids contain the bases A, C, and G; T, however, is found only in DNA, while U is found in RNA. The pentose sugar in DNA (2′-deoxyribose) differs from the sugar in RNA (ribose) by the absence of a hydroxyl group (−OH) on the 2′ carbon of the sugar ring. Without an attached phosphate group, the sugar attached to one of the bases is known as a nucleoside. The phosphate group connects successive sugar residues by bridging the 5′-hydroxyl group on one sugar to the 3′-hydroxyl group of the next sugar in the chain. These nucleoside linkages are called phosphodiester bonds and are the same in RNA and DNA.

Biosynthesis and Degradation

Nucleotides are synthesized from readily available precursors in the cell. The ribose phosphate portion of both purine and pyrimidine nucleotides is synthesized from glucose via the pentose phosphate pathway. The six-atom pyrimidine ring is synthesized first and subsequently attached to the ribose phosphate. The two rings in purines are synthesized while attached to the ribose phosphate during the assembly of adenine or guanine nucleosides. In both cases the end product is a nucleotide carrying a phosphate attached to the 5′ carbon on the sugar. Finally, a specialized enzyme called a kinase adds two phosphate groups using adenosine triphosphate (ATP) as the phosphate donor to form ribonucleoside triphosphate, the immediate precursor of RNA. For DNA, the 2′-hydroxyl group is removed from the ribonucleoside diphosphate to give deoxyribonucleoside diphosphate. An additional phosphate group from ATP is then added by another kinase to form a deoxyribonucleoside triphosphate, the immediate precursor of DNA.

During normal cell metabolism, RNA is constantly being made and broken down. The purine and pyrimidine residues are reused by several salvage pathways to make more genetic material. Purine is salvaged in the form of the corresponding nucleotide, whereas pyrimidine is salvaged as the nucleoside.

Deoxyribonucleic Acid (DNA)

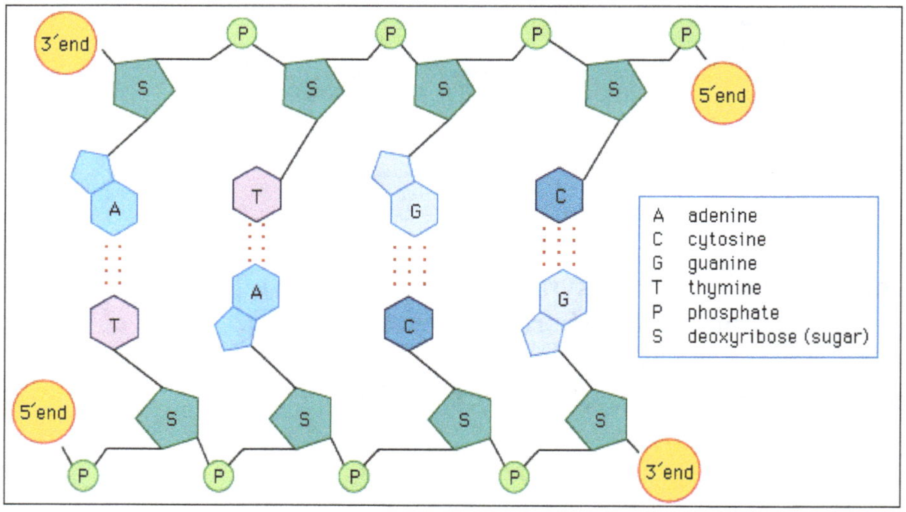

DNA is a polymer of the four nucleotides A, C, G, and T, which are joined through a backbone of alternating phosphate and deoxyribose sugar residues. These nitrogen-containing bases occur in complementary pairs as determined by their ability to form hydrogen bonds between them. A always pairs with T through two hydrogen bonds, and G always pairs with C through three hydrogen bonds. The spans of A:T and G:C hydrogen-bonded pairs are nearly identical, allowing them to bridge the sugar-phosphate chains uniformly. This structure, along with the molecule's chemical stability, makes DNA the ideal genetic material. The bonding between complementary bases also provides a mechanism for the replication of DNA and the transmission of genetic information.

DNA structure, showing the nucleotide bases cytosine (C), thymine (T), adenine (A), and guanine (G) linked to a backbone of alternating phosphate (P) and deoxyribose sugar (S) groups. Two sugar-phosphate chains are paired through hydrogen bonds between A and T and between G and C, thus forming the twin-stranded double helix of the DNA molecule.

Chemical Structure

In 1953 James D. Watson and Francis H.C. Crick proposed a three-dimensional structure for DNA based on low-resolution X-ray crystallographic data and on Erwin Chargaff's observation that, in naturally occurring DNA, the amount of T equals the amount of A and the amount of G equals the amount of C. Watson and Crick, who shared a Nobel Prize in 1962 for their efforts, postulated that two strands of polynucleotides coil around each other, forming a double helix. The two strands, though identical, run in opposite directions as determined by the orientation of the 5′ to 3′ phosphodiester bond. The sugar-phosphate chains run along the outside of the helix, and the bases lie on the inside, where they are linked to complementary bases on the other strand through hydrogen bonds.

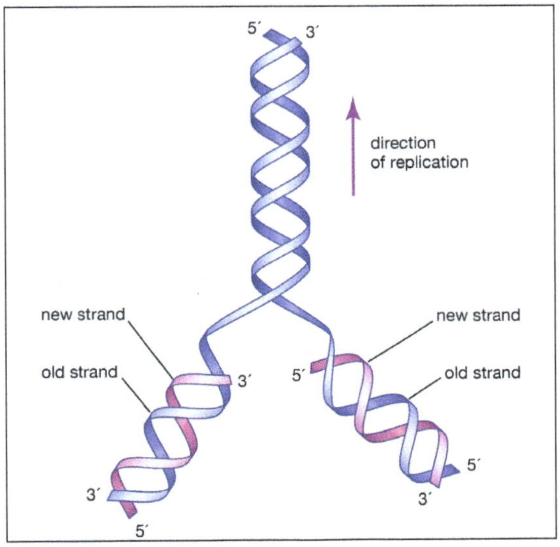

The initial proposal of the structure of DNA by James Watson and Francis Crick, which was accompanied by a suggestion on the means of replication.

The double helical structure of normal DNA takes a right-handed form called the B-helix. The helix makes one complete turn approximately every 10 base pairs. B-DNA has two principal grooves, a wide major groove and a narrow minor groove. Many proteins interact in the space of the major groove, where they make sequence-specific contacts with the bases. In addition, a few proteins are known to make contacts via the minor groove.

Several structural variants of DNA are known. In A-DNA, which forms under conditions of high salt concentration and minimal water, the base pairs are tilted and displaced toward the minor groove. Left-handed Z-DNA forms most readily in strands that contain sequences with alternating purines and pyrimidines. DNA can form triple helices when two strands containing runs of pyrimidines interact with a third strand containing a run of purines.

B-DNA is generally depicted as a smooth helix; however, specific sequences of bases can distort the otherwise regular structure. For example, short tracts of A residues interspersed with short sections of general sequence result in a bent DNA molecule. Inverted base sequences, on the other hand, produce cruciform structures with four-way junctions that are similar to recombination intermediates. Most of these alternative DNA structures have only been characterized in the laboratory, and their cellular significance is unknown.

Biological Structures

Naturally occurring DNA molecules can be circular or linear. The genomes of single-celled bacteria and archaea (the prokaryotes), as well as the genomes of mitochondria and chloroplasts (certain functional structures within the cell), are circular molecules. In addition, some bacteria and archaea have smaller circular DNA molecules called plasmids that typically contain only a few genes. Many plasmids are readily transmitted from one cell to another. For a typical bacterium, the genome that encodes all of the genes of the organism is a single contiguous circular molecule that contains a half million to five million base pairs. The genomes of most eukaryotes and some prokaryotes contain linear DNA molecules called chromosomes. Human DNA, for example, consists of 23 pairs of linear chromosomes containing three billion base pairs.

In all cells, DNA does not exist free in solution but rather as a protein-coated complex called chromatin. In prokaryotes, the loose coat of proteins on the DNA helps to shield the negative charge of the phosphodiester backbone. Chromatin also contains proteins that control gene expression and determine the characteristic shapes of chromosomes. In eukaryotes, a section of DNA between 140 and 200 base pairs long winds around a discrete set of eight positively charged proteins called a histone, forming a spherical structure called the nucleosome. Additional histones are wrapped by successive sections of DNA, forming a series of nucleosomes like beads on a string. Transcription and replication of DNA is more complicated in eukaryotes because the nucleosome complexes have to be at least partially disassembled for the processes to proceed effectively.

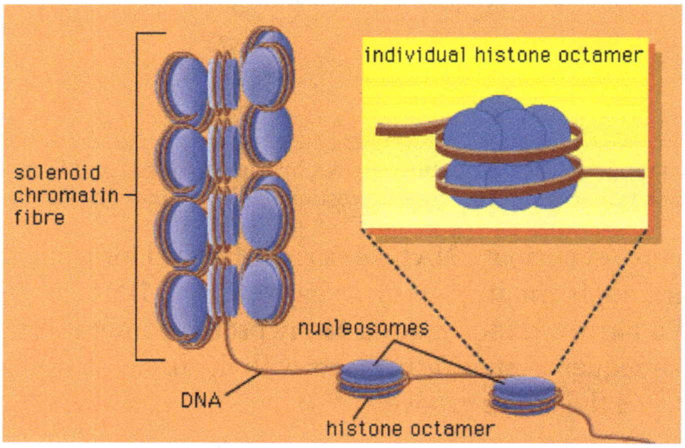

DNA wrapped around clusters of histone proteins to form nucleosomes, which are coiled to form solenoids, the basis of the chromatin fibre that makes up chromosomes.

Most prokaryote viruses contain linear genomes that typically are much shorter and contain only the genes necessary for viral propagation. Bacterial viruses called bacteriophages (or phages) may contain both linear and circular forms of DNA. For instance, the genome of bacteriophage λ (lambda), which infects the bacterium Escherichia coli, contains 48,502 base pairs and can exist as a linear molecule packaged in a protein coat. The DNA of phage λ can also exist in a circular form that is able to integrate into the circular genome of the host bacterial cell. Both circular and linear genomes are found among eukaryotic viruses, but they more commonly use RNA as the genetic material.

Biochemical Properties

Denaturation

The strands of the DNA double helix are held together by hydrogen bonding interactions between the complementary base pairs. Heating DNA in solution easily breaks these hydrogen bonds, allowing the two strands to separate—a process called denaturation or melting. The two strands may reassociate when the solution cools, reforming the starting DNA duplex—a process called renaturation or hybridization. These processes form the basis of many important techniques for manipulating DNA. For example, a short piece of DNA called an oligonucleotide can be used to test whether a very long DNA sequence has the complementary sequence of the oligonucleotide embedded within it. Using hybridization, a single-stranded DNA molecule can capture complementary sequences from any source. Single strands from RNA can also reassociate. DNA and RNA single strands can form hybrid molecules that are even more stable than double-stranded DNA. These molecules form the basis of a technique that is used to purify and characterize messenger RNA (mRNA) molecules corresponding to single genes.

Ultraviolet Absorption

DNA melting and reassociation can be monitored by measuring the absorption of ultraviolet (UV) light at a wavelength of 260 nanometres (billionths of a metre). When DNA is in a double-stranded conformation, absorption is fairly weak, but when DNA is single-stranded, the unstacking of the bases leads to an enhancement of absorption called hyperchromicity. Therefore, the extent to which DNA is single-stranded or double-stranded can be determined by monitoring UV absorption.

Chemical Modification

After a DNA molecule has been assembled, it may be chemically modified—sometimes deliberately by special enzymes called DNA methyltransferases and sometimes accidentally by oxidation, ionizing radiation, or the action of chemical carcinogens. DNA can also be cleaved and degraded by enzymes called nucleases.

Methylation

Three types of natural methylation have been reported in DNA. Cytosine can be modified either on the ring to form 5-methylcytosine or on the exocyclic amino group to form N^4-methylcytosine. Adenine may be modified to form N6-methyladenine. N^4-methylcytosine and N^6-methyladenine are found only in bacteria and archaea, whereas 5-methylcytosine is widely distributed. Special enzymes called DNA methyltransferases are responsible for this methylation; they recognize specific sequences within the DNA molecule so that only a subset of the bases is modified. Other methylations of the bases or of the deoxyribose are sometimes induced by carcinogens. These usually lead to mispairing of the bases during replication and have to be removed if they are not to become mutagenic.

Natural methylation has many cellular functions. In bacteria and archaea, methylation forms an essential part of the immune system by protecting DNA molecules from fragmentation by restriction endonucleases. In some organisms, methylation helps to eliminate incorrect base sequences introduced during DNA replication. By marking the parental strand with a methyl group, a cellular mechanism known as the mismatch repair system distinguishes between the newly replicated strand where the errors occur and the correct sequence on the template strand. In higher eukaryotes, 5-methylcytosine controls many cellular phenomena by preventing DNA transcription. Methylation is also believed to signal imprinting, a process whereby some genes inherited from one parent are selectively inactivated. Correct methylation may also repress or activate key genes that control embryonic development. On the other hand, 5-methylcytosine is potentially mutagenic because thymine produced during the methylation process converts C:G pairs to T:A pairs. In mammals, methylation takes place selectively within the dinucleotide sequence CG—a rare sequence, presumably because it has been lost by mutation. In many cancers, mutations are found in key genes at CG dinucleotides.

Nucleases

Nucleases are enzymes that hydrolytically cleave the phosphodiester backbone of DNA. Endonucleases cleave in the middle of chains, while exonucleases operate selectively by degrading from the end of the chain. Nucleases that act on both single- and double-stranded DNA are known.

Restriction endonucleases are a special class that recognize and cleave specific sequences in DNA. Type II restriction endonucleases always cleave at or near their recognition sites. They produce small, well-defined fragments of DNA that help to characterize genes and genomes and that produce recombinant DNAs. Fragments of DNA produced by restriction endonucleases can be moved from one organism to another. In this way it has been possible to express proteins such as human insulin in bacteria.

Mutation

Chemical modification of DNA can lead to mutations in the genetic material. Anions such as bisulfite can deaminate cytosine to form uracil, changing the genetic message by causing C-to-T transitions. Exposure to acid causes the loss of purine residues, though specific enzymes exist in cells to repair these lesions. Exposure to UV light can cause adjacent pyrimidines to dimerize, while oxidative damage from free radicals or strong oxidizing agents can cause a variety of lesions that are mutagenic if not repaired. Halogens such as chlorine and bromine react directly with uracil, adenine, and guanine, giving substituted bases that are often mutagenic. Similarly, nitrous acid reacts with primary amine groups—for example, converting adenosine into inosine—which then leads to changes in base pairing and mutation. Many chemical mutagens, such as chlorinated hydrocarbons and nitrites, owe their toxicity to the production of halides and nitrous acid during their metabolism in the body.

Supercoiling

Circular DNA molecules such as those found in plasmids or bacterial chromosomes can adopt many different topologies. One is active supercoiling, which involves the cleavage of one DNA strand, its winding one or more turns around the complementary strand, and then the resealing of the molecule. Each complete rotation leads to the introduction of one supercoiled turn in the DNA, a process that can continue until the DNA is fully wound and collapses on itself in a tight ball. Reversal is also possible. Special enzymes called gyrases and topoisomerases catalyze the winding and relaxation of supercoiled DNA. In the linear chromosomes of eukaryotes, the DNA is usually tightly constrained at various points by proteins, allowing the intervening stretches to be supercoiled. This property is partially responsible for the great compaction of DNA that is necessary to fit it within the confines of the cell. The DNA in one human cell would have an extended length of between two and three metres, but it is packed very tightly so that it can fit within a human cell nucleus that is 10 micrometres in diameter.

Sequence Determination

Methods to determine the sequences of bases in DNA were pioneered in the 1970s by Frederick Sanger and Walter Gilbert, whose efforts won them a Nobel Prize in 1980. The Gilbert-Maxam method relies on the different chemical reactivities of the bases, while the Sanger method is based on enzymatic synthesis of DNA in vitro. Both methods measure the distance from a fixed point on DNA to each occurrence of a particular base—A, C, G, or T. DNA fragments obtained from a series of reactions are separated according to length in four "lanes" by gel electrophoresis. Each lane corresponds to a unique base, and the sequence is read directly from the gel. The Sanger method has now been automated using fluorescent dyes to label the DNA, and a single machine can produce tens of thousands of DNA base sequences in a single run.

Ribonucleic Acid (RNA)

RNA is a single-stranded nucleic acid polymer of the four nucleotides A, C, G, and U joined through a backbone of alternating phosphate and ribose sugar residues. It is the first intermediate in converting the information from DNA into proteins essential for the working of a cell. Some RNAs also serve direct roles in cellular metabolism. RNA is made by copying the base sequence of a section of double-stranded DNA, called a gene, into a piece of single-stranded nucleic acid. This process, called transcription, is catalyzed by an enzyme called RNA polymerase.

Chemical Structure

Whereas DNA provides the genetic information for the cell and is inherently quite stable, RNA has many roles and is much more reactive chemically. RNA is sensitive to oxidizing agents such as periodate that lead to opening of the 3′-terminal ribose ring. The 2′-hydroxyl group on the ribose ring is a major cause of instability in RNA, because the presence of alkali leads to rapid cleavage of the phosphodiester bond linking ribose and phosphate groups. In general, this instability is not a significant problem for the cell, because RNA is constantly being synthesized and degraded.

Interactions between the nitrogen-containing bases differ in DNA and RNA. In DNA, which is usually double-stranded, the bases in one strand pair with complementary bases in a second DNA strand. In RNA, which is usually single-stranded, the bases pair with other bases within the same molecule, leading to complex three-dimensional structures. Occasionally, intermolecular RNA/RNA duplexes do form, but they form a right-handed A-type helix rather than the B-type DNA helix. Depending on the amount of salt present, either 11 or 12 base pairs are found in each turn of the helix. Helices between RNA and DNA molecules also form; these adopt the A-type conformation and are more stable than either RNA/RNA or DNA/DNA duplexes. Such hybrid duplexes are important species in biology, being formed when RNA polymerase transcribes DNA into mRNA for protein synthesis and when reverse transcriptase copies a viral RNA genome such as that of the human immunodeficiency virus (HIV).

Single-stranded RNAs are flexible molecules that form a variety of structures through internal base pairing and additional non-base pair interactions. They can form hairpin loops such as those found in transfer RNA (tRNA), as well as longer-range interactions involving both the bases and the phosphate residues of two or more nucleotides. This leads to compact three-dimensional structures. Most of these structures have been inferred from biochemical data, since few crystallographic images are available for RNA molecules. In some types of RNA, a large number of bases are modified after the RNA is transcribed. More than 90 different modifications have been documented, including extensive methylations and a wide variety of substitutions around the ring. In some cases these modifications are known to affect structure and are essential for function.

Types of RNA

Messenger RNA (mRNA)

Messenger RNA (mRNA) delivers the information encoded in one or more genes from the DNA to the ribosome, a specialized structure, or organelle, where that information is decoded into a protein. In prokaryotes, mRNAs contain an exact transcribed copy of the original DNA sequence with a terminal 5′-triphosphate group and a 3′-hydroxyl residue. In eukaryotes the mRNA molecules are more elaborate. The 5′-triphosphate residue is further esterified, forming a structure called a cap. At the 3′ ends, eukaryotic mRNAs typically contain long runs of adenosine residues (polyA) that are not encoded in the DNA but are added enzymatically after transcription. Eukaryotic mRNA molecules are usually composed of small segments of the original gene and are generated by a process of cleavage and rejoining from an original precursor RNA (pre-mRNA) molecule, which is an exact copy of the gene. In general, prokaryotic mRNAs are degraded very rapidly, whereas the cap structure and the polyA tail of eukaryotic mRNAs greatly enhance their stability.

Ribosomal RNA (rRNA)

Ribosomal RNA (rRNA) molecules are the structural components of the ribosome. The rRNAs form extensive secondary structures and play an active role in recognizing conserved portions of mRNAs and tRNAs. They also assist with the catalysis of protein synthesis. In the prokaryote E. coli, seven copies of the rRNA genes synthesize about 15,000 ribosomes per cell. In eukaryotes the numbers are much larger. Anywhere from 50 to 5,000 sets of rRNA genes and as many as 10 million ribosomes may be present in a single cell. In eukaryotes these rRNA genes are looped out of the main chromosomal fibres and coalesce in the presence of proteins to form an organelle called the nucleolus. The nucleolus is where the rRNA genes are transcribed and the early assembly of ribosomes takes place.

Transfer RNA (tRNA)

Transfer RNA (tRNA) carries individual amino acids into the ribosome for assembly into the growing polypeptide chain. The tRNA molecules contain 70 to 80 nucleotides and fold into a characteristic cloverleaf structure. Specialized tRNAs exist for each of the 20 amino acids needed for protein synthesis, and in many cases more than one tRNA for each amino acid is present. The nucleotide sequence is converted into a protein sequence by translating each three-base sequence (called a codon) with a specific protein. The 61 codons used to code amino acids can be read by many fewer than 61 distinct tRNAs. In E. coli a total of 40 different tRNAs are used to translate the

61 codons. The amino acids are loaded onto the tRNAs by specialized enzymes called aminoacyl tRNA synthetases, usually with one synthetase for each amino acid. However, in some organisms, less than the full complement of 20 synthetases are required because some amino acids, such as glutamine and asparagine, can be synthesized on their respective tRNAs. All tRNAs adopt similar structures because they all have to interact with the same sites on the ribosome.

Ribozymes

Not all catalysis within the cell is carried out exclusively by proteins. Thomas Cech and Sidney Altman, jointly awarded a Nobel Prize in 1989, discovered that certain RNAs, now known as ribozymes, showed enzymatic activity. Cech showed that a noncoding sequence (intron) in the small subunit rRNA of protozoans, which had to be removed before the rRNA was functional, can excise itself from a much longer precursor RNA molecule and rejoin the two ends in an autocatalytic reaction. Altman showed that the RNA component of an RNA protein complex called ribonuclease P can cleave a precursor tRNA to generate a mature tRNA. In addition to self-splicing RNAs similar to the one discovered by Cech, artificial RNAs have been made that show a variety of catalytic reactions. It is now widely held that there was a stage during evolution when only RNA catalyzed and stored genetic information. This period, sometimes called "the RNA world," is believed to have preceded the function of DNA as genetic material.

Antisense RNAs

Most antisense RNAs are synthetically modified derivatives of RNA or DNA with potential therapeutic value. In nature, antisense RNAs contain sequences that are the complement of the normal coding sequences found in mRNAs (also called sense RNAs). Like mRNAs, antisense RNAs are single-stranded, but they cannot be translated into protein. They can inactivate their complementary mRNA by forming a double-stranded structure that blocks the translation of the base sequence. Artificially introducing antisense RNAs into cells selectively inactivates genes by interfering with normal RNA metabolism.

Viral Genomes

Many viruses use RNA for their genetic material. This is most prevalent among eukaryotic viruses, but a few prokaryotic RNA viruses are also known. Some common examples include poliovirus, human immunodeficiency virus (HIV), and influenza virus, all of which affect humans, and tobacco mosaic virus, which infects plants. In some viruses the entire genetic material is encoded in a single RNA molecule, while in the segmented RNA viruses several RNA molecules may be present. Many RNA viruses such as HIV use a specialized enzyme called reverse transcriptase that permits replication of the virus through a DNA intermediate. In some cases this DNA intermediate becomes integrated into the host chromosome during infection; the virus then exists in a dormant state and effectively evades the host immune system.

Other RNAs

Many other small RNA molecules with specialized functions are present in cells. For example, small nuclear RNAs (snRNAs) are involved in RNA splicing, and other small RNAs that form part of the enzymes telomerase or ribonuclease P are part of ribonucleoprotein particles. The RNA component of telomerase contains a short sequence that serves as a template for the addition of small strings of oligonucleotides at the ends of eukaryotic chromosomes. Other RNA molecules serve as guide RNAs for editing, or they are complementary to small sections of rRNA and either direct the positions at which methyl groups need to be added or mark U residues for conversion to the isomer pseudouridine.

RNA Processing

Cleavage

Following synthesis by transcription, most RNA molecules are processed before reaching their final form. Many rRNA molecules are cleaved from much larger transcripts and may also be methylated or enzymatically modified. In addition, tRNAs are usually formed as longer precursor molecules that are cleaved by ribonuclease P to generate the mature 5′ end and often have extra residues added to their 3′ end to form the sequence CCA. The hydroxyl group on the ribose ring of the terminal A of the 3′-CCA sequence acts as the amino acid acceptor necessary for the function of RNA in protein building.

Splicing

In prokaryotes the protein coding sequence occupies one continuous linear segment of DNA. However, in eukaryotic genes the coding sequences are frequently "split" in the genome—a discovery reached independently in the 1970s by Richard J. Roberts and Phillip A. Sharp, whose work won them a Nobel Prize in 1993. The segments of DNA or RNA coding for protein are called exons, and the noncoding regions separating the exons are called introns. Following transcription, these coding sequences must be joined together before the mRNAs can function. The process of removal of the introns and subsequent rejoining of the exons is called RNA splicing. Each intron is removed in a separate series of reactions by a complicated piece of enzymatic machinery called a spliceosome. This machinery consists of a number of small nuclear ribonucleoprotein particles (snRNPs) that contain small nuclear RNAs (snRNAs).

RNA Editing

Some RNA molecules, particularly those in protozoan mitochondria, undergo extensive editing following their initial synthesis. During this editing process, residues are added or deleted by a posttranscriptional mechanism under the influence of guide RNAs. In some cases as much as 40 percent of the final RNA molecule may be derived by this editing process, rather than being coded directly in the genome. Some examples of editing have also been found in mRNA molecules, but these appear much more limited in scope.

Nucleic Acid Metabolism

DNA Metabolism

Replication, repair, and recombination—the three main processes of DNA metabolism—are carried out by specialized machinery within the cell. DNA must be replicated accurately in order to ensure the integrity of the genetic code. Errors that creep in during replication or because of damage after replication must be repaired. Finally, recombination between genomes is an important mechanism to provide variation within a species and to assist the repair of damaged DNA. The details of each process have been worked out in prokaryotes, where the machinery is more streamlined, simpler, and more amenable to study. Many of the basic principles appear to be similar in eukaryotes.

Replication

Basic Mechanisms

DNA replication is a semiconservative process in which the two strands are separated and new complementary strands are generated independently, resulting in two exact copies of the original DNA molecule. Each copy thus contains one strand that is derived from the parent and one newly synthesized strand. Replication begins at a specific point on a chromosome called an origin, proceeds in both directions along the strand, and ends at a precise point. In the case of circular chromosomes, the end is reached automatically when the two extending chains meet, at which point specific proteins join the strands. DNA polymerases cannot initiate replication at the end of a DNA strand; they can only extend preexisting oligonucleotide fragments called primers. Therefore, in linear chromosomes, special mechanisms initiate and terminate DNA synthesis to avoid loss of information. The initiation of DNA synthesis is usually preceded by synthesis of a short RNA primer by a specialized RNA polymerase called primase. Following DNA replication, the initiating primer RNAs are degraded.

The two DNA strands are replicated in different fashions dictated by the direction of the phosphodiester bond. The leading strand is replicated continuously by adding individual nucleotides to the 3′ end of the chain. The lagging strand is synthesized in a discontinuous manner by laying down short RNA primers and then filling the gaps by DNA polymerase, such that the bases are always added in the 5′ to 3′ direction. The short RNA fragments made during the copying of the lagging strand are degraded when no longer needed. The two newly synthesized DNA segments are joined by an enzyme called DNA ligase. In this way, replication can proceed in both directions, with two leading strands and two lagging strands proceeding outward from the origin.

Enzymes of Replication

DNA polymerase adds single nucleotides to the 3′ end of either an RNA or a DNA molecule. In the prokaryote E. coli, there are three DNA polymerases; one is responsible for chromosome replication, and the other two are involved in the resynthesis of DNA during damage repair. DNA polymerases of eukaryotes are even more complicated. In human cells, for instance, more than five different DNA polymerases have been characterized. Separate polymerases catalyze the synthesis of the leading and lagging strands in human cells, and a separate polymerase is responsible for replication of mitochondrial DNA. The other polymerases are involved in the repair of DNA damage.

A number of other proteins are also essential for replication. Proteins called DNA helicases help to separate the two strands of DNA, and single-stranded DNA binding proteins stabilize them during opening prior to being copied. The opening of the DNA helix introduces considerable strain in the form of supercoiling, a movement that is subsequently relaxed by enzymes called topoisomerases. A special RNA polymerase called primase synthesizes the primers needed at the origin to begin transcription, and DNA ligase seals the nicks formed between individual fragments.

The ends of linear eukaryotic chromosomes are marked by special sequences called telomeres that are synthesized by a special DNA polymerase called telomerase. This enzyme contains an RNA component that serves as a template for the exact sequence found at the ends of chromosomes. Multiple copies of a short sequence within the telomerase-associated RNA are made and added to the telomere ends. This has the effect of preventing shortening of the DNA chain that would otherwise occur during replication.

Single-stranded viral genomes, mitochondrial genomes, and some viral genomes are replicated in specialized ways. Several viruses such as adenovirus use a nucleotide covalently bound to a protein as a primer, and the protein remains covalently bound to the DNA after replication. Many single-stranded viruses use a rolling circle mechanism of replication whereby a double-stranded copy of the virus is first made. The replicating machinery then copies the nonviral strand in a continuous fashion, generating long single-stranded DNA from which full-length viral DNA strands are excised by specialized nucleases.

Recombination

Recombination is the principal mechanism through which variation is introduced into populations. For example, during meiosis, the process that produces sex cells (sperm or eggs), homologous chromosomes—one derived from the mother and the equivalent from the father—become paired, and recombination, or crossing-over, takes place. The two DNA molecules are fragmented, and similar segments of the chromosome are shuffled to produce two new chromosomes, each being a mosaic of the originals. The pair separates so that each sperm or egg receives just one of the shuffled chromosomes. When sperm and egg fuse, the normal set of two copies of each chromosome is restored.

There are two forms of recombination, general and site-specific. General recombination typically involves cleavage and rejoining at identical or very similar sequences. In site-specific recombination, cleavage takes place at a specific site into which DNA is usually inserted. General recombination occurs among viruses during infection, in bacteria during conjugation, during transformation whereby DNA is directly introduced into cells, and during some types of repair processes. Site-specific recombination is frequently involved in the parasitic distribution of DNA segments throughout genomes. Many viruses, as well as special segments of DNA called transposons, rely on site-specific recombination to multiply and spread.

General Recombination

General recombination, also called homologous recombination, involves two DNA molecules that have long stretches of similar base sequences. The DNA molecules are nicked to produce single strands; these subsequently invade the other duplex, where base pairing leads to a four-stranded DNA structure. The cruciform junction within this structure is called a Holliday junction, named after Robin Holliday, who proposed the original model for homologous recombination in 1964. The Holliday junction travels along the DNA duplex by "unzipping" one strand and reforming the hydrogen bonds on the second strand. Following this branch migration, the two duplexes can be nicked again, allowing them to separate. Finally, the nicks are repaired by DNA ligase. The result is two DNA duplexes in which the segment between the two nicks has been replaced. The enzymes involved in recombination have been characterized best in the prokaryote E. coli. A key enzyme is

RecA, which catalyzes the strand invasion process. RecA coats single-stranded DNA and facilitates its pairing with a double-stranded DNA molecule containing the same sequence, which produces a loop structure.

Another protein, known as RecBC, is important for the recombination process. Functioning at free ends of DNA, RecBC catalyzes an unwinding-rewinding reaction as it traverses the length of the molecule. Since unwinding is faster than rewinding, a loop is produced behind the enzyme that facilitates subsequent pairing with another DNA molecule. A number of other proteins are also important for recombination, including single-stranded DNA binding proteins that stabilize single-stranded DNA, DNA polymerase to repair any gaps that might be formed, and DNA ligase to reseal the nicks after recombination is complete. The details of eukaryotic recombination are expected to parallel those found in E. coli, although the highly compact chromatin structure in eukaryotes makes the process more complicated.

It is important to note that the initial product of recombination between two regions of DNA that are similar but not identical will be a "heteroduplex"—that is, a molecule in which mismatched bases will be present at some positions in the helix. Thus, in the specialized recombination that takes place during meiosis, one round of replication is necessary before the mosaic chromosomes produced by recombination are properly matched. Enzymes are present in cells that specifically recognize and repair mismatches, so that the initial products of recombination can sometimes be repaired before they are replicated. In such cases the final products of replication will not be true reciprocal events, but rather one of the original parental molecules will appear to have been maintained to the exclusion of the other—a process called gene conversion.

Recombination also functions occasionally to repair lesions in DNA. If one chromosome of a pair becomes irreversibly damaged, the information from the other chromosome can be copied and inserted by recombination to provide a correct replacement of the damaged section. The key idea here is that sequences flanking the damage from a sister chromosome can base-pair with the corresponding sequences on the damaged chromosome, thus allowing replication to copy the correct sequence and repair the lesion.

Site-specific Recombination

Site-specific recombination involves very short specific sequences that are recognized by proteins. Long DNA sequences such as viral genomes, drug-resistance elements, and regulatory sequences such as the mating type locus in yeast can be inserted, removed, or inverted, having profound regulatory effects. More than any other mechanism, site-specific recombination is responsible for reshaping genomes. For example, the genomes of many higher organisms, including plants and humans, show evidence that transposable elements have been constantly inserted throughout the genome and even into one another from time to time.

One example of site-specific recombination is the integration of DNA from bacteriophage λ into the chromosome of E. coli. In this reaction, bacteriophage λ DNA, which is a linear molecule in the normal phage, first forms a circle and then is cleaved by the enzyme λ-integrase at a specific site called the phage attachment site. A similar site on the bacterial chromosome is cut by integrase to give ends with the identical extension. Because of the complementarity between these two ends, they can be rejoined so that the original circular λ chromosome is inserted into the chromosome of the E. coli bacterium. Once integrated, the phage can be held in an inactive state until signals are generated that reverse the process, allowing the phage genome to escape and resume its normal life cycle of growth and spread into other bacteria. This site-specific recombination process requires only λ-integrase and one host DNA binding protein called the integration host factor. A third protein, called excisionase, recognizes the hybrid sites formed on integration and, in conjunction with integrase, catalyzes an excision process whereby the λ chromosome is removed from the bacterial chromosome.

A similar but more widespread version of DNA integration and excision is exhibited by the transposons, the so-called jumping genes. These elements range in size from fewer than 1,000 to as many as 40,000 base pairs. Transposons are able to move from one location in a genome to another, as first discovered in corn (maize) during the 1940s and '50s by Barbara McClintock, whose work won her a Nobel Prize in 1983. Most, if not all, transposons encode an enzyme called transposase that acts much like λ-integrase by cleaving the ends of the transposon as well as its target site. Transposons differ from bacteriophage λ in that they do not have a separate existence outside of the chromosome but rather are always maintained in an integrated site. Two types of transposition can occur—one in which the element simply moves from one site in the chromosome to another and a second in which the transposon is replicated prior to moving. This second type of transposition leaves behind the original copy of the transposon and generates a second copy that is inserted elsewhere in the genome. Known as replicative transposition, this process is the mechanism responsible for the vast spread of transposable elements in many higher organisms.

The simplest kinds of transposons merely contain a copy of the transposase with no additional genes. They behave as parasitic elements and usually have no known associated function that is advantageous to the host. More often, transposable elements have additional genes associated with them—for example, antibiotic resistance factors. Antibiotic resistance typically occurs when an infecting bacterium acquires a plasmid that carries a gene encoding resistance to one or more antibiotics. Typically, these resistance genes are carried on transposable elements that have moved into plasmids and are easily transferred from one organism to another. Once a bacterium picks up such a gene, it enjoys a great selective advantage because it can grow in the presence of the antibiotic. Indiscriminate use of antibiotics actually promotes the buildup of these drug-resistant plasmids and strains.

Repair

It is extremely important that the integrity of DNA be maintained in order to ensure the accurate workings of a cell over its lifetime and to make certain that genetic information is accurately passed from one generation to the next. This maintenance is achieved by repair processes that constantly monitor the DNA for lesions and activate appropriate repair enzymes. Serious lesions in DNA such as pyrimidine dimers or gaps can be repaired by recombination mechanisms, but there are many other repair mechanisms.

One important mechanism is that of mismatch repair, which has been studied extensively in E. coli. The system is directed by the presence of a methyl group within the sequence GATC on the template strand. Comparable systems for mismatch repair also operate in eukaryotes, though the template strand is not marked by methyl groups. In fact, lesions within the genes for human mismatch repair systems are known to be responsible for many cancers. Loss of the mismatch repair system allows mutations to build up quickly and eventually to affect the genes that cause cells to divide. As a result, cells divide in an uncontrolled manner and become cancerous.

Once replication is complete, the most common kind of damage to nucleic acids is one in which the normal A, C, G, and T bases are changed into chemically modified bases that usually differ significantly from their natural counterparts. The only exceptions are the deamination of cytosine to uracil and the deamination of 5-methylcytosine to thymine. In these cases the product is a G:U or G:T mismatch. Specific enzymes called DNA glycosylases can recognize uracil in DNA or the thymine in a G:T mismatch and can selectively remove the base by cleaving the bond between the base and the deoxyribose sugar. Many of these enzymes are specific for the different chemically modified bases that may be present in DNA.

Another common means of repairing DNA lesions is by an excision repair pathway. Enzymes recognize damage within DNA, probably by detecting an altered conformation of DNA, and then nick the strand on either side of the lesion, allowing a small single-stranded DNA to be excised. DNA polymerase and DNA ligase then repair the single-stranded gap. In all of these systems, the presence of an abnormal base signifies which strand is to be repaired, and the complementary strand is used as the template to ensure the accuracy of repair.

RNA Metabolism

RNA provides the link between the genetic information encoded in DNA and the actual workings of the cell. Some RNA molecules such as the rRNAs and the snRNAs become part of complicated ribonucleoprotein structures with specialized roles in the cell. Others such as tRNAs play key roles in protein synthesis, while mRNAs direct the synthesis of proteins by the ribosome. Three distinct phases of RNA metabolism occur. First, selected segments of the genome are copied by transcription to produce the precursor RNAs. Second, these precursors are processed to become functionally mature RNAs ready for use. When these RNAs are mRNAs, they are then used for translation. Third, after use the RNAs are degraded, and the bases are recycled. Thus, transcription is the process where a specific segment of DNA, a gene, is copied into a specific RNA that encodes a single protein or plays a structural or catalytic role. Translation is the decoding of the information within mRNA molecules that takes place on a specialized structure called a ribosome. There are important differences in both transcription and translation between prokaryotic and eukaryotic organisms.

Transcription

Small segments of DNA are transcribed into RNA by the enzyme RNA polymerase, which achieves this copying in a strictly controlled process. The first step is to recognize a specific sequence on DNA called a promoter that signifies the start of the gene. The two strands of DNA become separated at this point, and RNA polymerase begins copying from a specific point on one strand of the DNA using a ribonucleoside 5′-triphosphate to begin the growing chain. Additional ribonucleoside triphosphates are used as the substrate, and, by cleavage of their high-energy phosphate bond, ribonucleoside monophosphates are incorporated into the growing RNA chain. Each successive ribonucleotide is directed by the complementary base pairing rules of DNA. Thus, a C in DNA directs the incorporation of a G into RNA, G is copied into C, T into A, and A into U. Synthesis continues until a termination signal is reached, at which point the RNA polymerase drops off the

DNA, and the RNA molecule is released. In some cases this RNA molecule is the final mRNA. In other cases it is a pre-mRNA and requires further processing before it is ready for translation by the ribosome. Ahead of many genes in prokaryotes, there are signals called "operators" where specialized proteins called repressors bind to the DNA just upstream of the start point of transcription and prevent access to the DNA by RNA polymerase. These repressor proteins thus prevent transcription of the gene by physically blocking the action of the RNA polymerase. Typically, repressors are released from their blocking action when they receive signals from other molecules in the cell indicating that the gene needs to be expressed. Ahead of some prokaryotic genes are signals to which activator proteins bind that positively induce transcription.

Transcription in higher organisms is more complicated. First, the RNA polymerase of eukaryotes is a more complicated enzyme than the relatively simple five-subunit enzyme of prokaryotes. In addition, there are many more accessory factors that help to control the efficiency of the individual promoters. These accessory proteins are called transcription factors and typically respond to signals from within the cell that indicate whether transcription is required. In many human genes, several transcription factors may be needed before transcription can proceed efficiently. A transcription factor can cause either repression or activation of gene expression in eukaryotes.

During transcription, only one strand of the DNA is usually copied. This is called the template strand, and the RNA molecules produced are single-stranded. The DNA strand that would correspond to the mRNA is called the coding or sense strand, and it is not unusual for this to change from one gene to the next. In eukaryotes the initial product of transcription is called a pre-mRNA, which is extensively spliced before the mature mRNA is produced, ready for translation by the ribosome.

Translation

The process of translation uses the information present in the nucleotide sequence of mRNA to direct the synthesis of a specific protein for use by the cell. Translation takes place on the ribosomes—complex particles in the cell that contain RNA and protein. In prokaryotes the ribosomes are loaded onto the mRNA while transcription is still ongoing. Near the 5′ end of the mRNA, a short sequence of nucleotides signals the starting point for translation. It contains a few nucleotides called a ribosome binding site, or Shine-Dalgarno sequence. In E. coli the tetranucleotide GAGG is sufficient to serve as a binding site. This typically lies five to eight bases upstream of an initiation codon. The mRNA sequence is read three bases at a time from its 5′ end toward its 3′ end, and one amino acid is added to the growing chain from its respective aminoacyl tRNA, until the complete protein chain is assembled. Translation stops when the ribosome encounters a termination codon, normally UAG, UAA, or UGA. Special release factors associate with the ribosome in response to these codons, and the newly synthesized protein, tRNAs, and mRNA all dissociate. The ribosome then becomes available to interact with another mRNA molecule.

In eukaryotes the essence of protein synthesis is the same, although the ribosomes are more complicated. As with prokaryotic initiation, the signal sequence interacts with the 3′ end of the small subunit rRNA during formation of the initiation complex.

The issue of fidelity is important during protein synthesis, but it is not as crucial as fidelity during replication. One mRNA molecule can be translated repeatedly to give many copies of the protein.

When an occasional protein is mistranslated, it usually does not fold properly and is then degraded by the cellular machinery. However, proofreading mechanisms exist within the ribosome to ensure accurate pairing between the codon in the mRNA and the anticodon in the tRNA.

One of the crowning achievements of molecular biology was the elucidation during the 1960s of the genetic code. Principals in this effort were Har G. Khorana and Marshall W. Nirenberg, who shared a Nobel Prize in 1968. Khorana and Nirenberg used artificial templates and protein synthesizing systems in the test tube to determine the coding potential of all 64 possible triplet codons.

The key feature of the genetic code is that the 20 amino acids are encoded by 61 codons. Thus, there is degeneracy in the code such that one amino acid is often specified by more than one codon. In the case of serine and leucine, six codons can be used for each. Among organisms that have been examined in detail, the code appears to be almost universal, from bacteria through archaea to eukaryotes. The known exceptions are found in the mitochondria of humans and many other organisms as well as in some species of bacteria. The structure within the genetic code whereby many amino acids are uniquely coded by the first two bases of the codon strongly suggests that the code has itself evolved from a more primitive code involving 16 dinucleotides. How the individual amino acids became associated with the different codons remains a matter of speculation.

CHROMOSOME

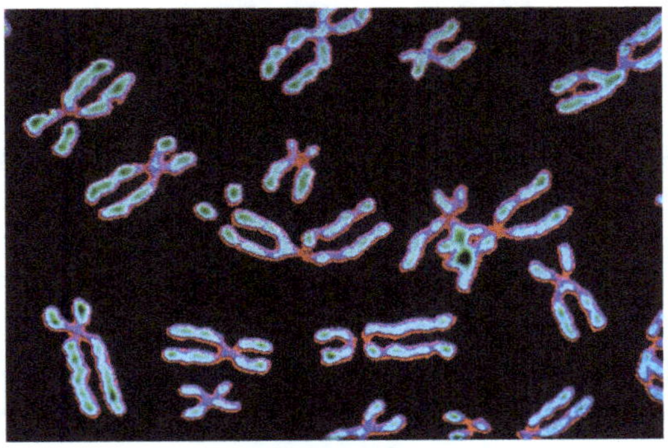

Human chromosomes.

Chromosome is the microscopic threadlike part of the cell that carries hereditary information in the form of genes. A defining feature of any chromosome is its compactness. For instance, the 46 chromosomes found in human cells have a combined length of 200 nm (1 nm = 10^{-9} metre); if the chromosomes were to be unraveled, the genetic material they contain would measure roughly 2 metres (about 6.5 feet) in length. The compactness of chromosomes plays an important role in helping to organize genetic material during cell division and enabling it to fit inside structures such as the nucleus of a cell, the average diameter of which is about 5 to 10 μm (1 μm = 0.001 mm, or 0.000039 inch), or the polygonal head of a virus particle, which may be in the range of just 20 to 30 nm in diameter.

The structure and location of chromosomes are among the chief differences between viruses, prokaryotes, and eukaryotes. The nonliving viruses have chromosomes consisting of either DNA (deoxyribonucleic acid) or RNA (ribonucleic acid); this material is very tightly packed into the viral head. Among organisms with prokaryotic cells (i.e., bacteria and blue-green algae), chromosomes consist entirely of DNA. The single chromosome of a prokaryotic cell is not enclosed within a nuclear membrane. Among eukaryotes, the chromosomes are contained in a membrane-bound cell nucleus. The chromosomes of a eukaryotic cell consist primarily of DNA attached to a protein core. They also contain RNA.

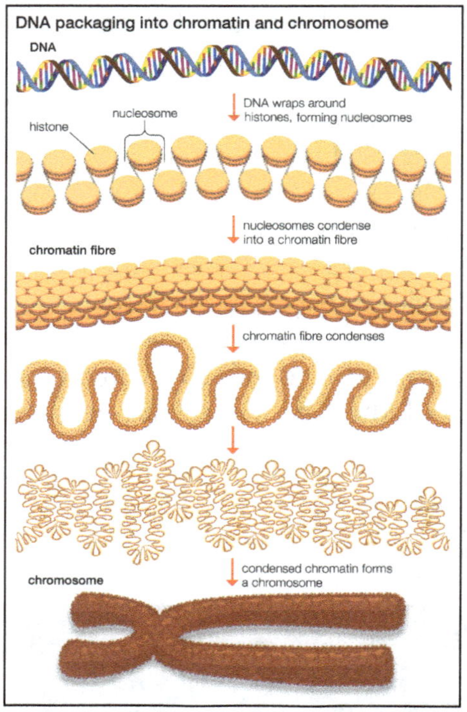

DNA wraps around proteins called histones to form units known as nucleosomes. These units condense into a chromatin fibre, which condenses further to form a chromosome.

Every eukaryotic species has a characteristic number of chromosomes (chromosome number). In species that reproduce asexually, the chromosome number is the same in all the cells of the organism. Among sexually reproducing organisms, the number of chromosomes in the body (somatic) cells is diploid (2n; a pair of each chromosome), twice the haploid (1n) number found in the sex cells, or gametes. The haploid number is produced during meiosis. During fertilization, two gametes combine to produce a zygote, a single cell with a diploid set of chromosomes.

Somatic cells reproduce by dividing, a process called mitosis. Between cell divisions the chromosomes exist in an uncoiled state, producing a diffuse mass of genetic material known as chromatin. The uncoiling of chromosomes enables DNA synthesis to begin. During this phase, DNA duplicates itself in preparation for cell division.

Following replication, the DNA condenses into chromosomes. At this point, each chromosome actually consists of a set of duplicate chromatids that are held together by the centromere. The centromere is the point of attachment of the kinetochore, a protein structure that is connected to the spindle fibres (part of a structure that pulls the chromatids to opposite ends of the cell). During the middle stage in cell division, the centromere duplicates, and the chromatid pair separates; each chromatid becomes a separate chromosome at this point. The cell divides, and both of the daughter cells have a complete (diploid) set of chromosomes. The chromosomes uncoil in the new cells, again forming the diffuse network of chromatin.

Among many organisms that have separate sexes, there are two basic types of chromosomes: sex chromosomes and autosomes. Autosomes control the inheritance of all the characteristics except the sex-linked ones, which are controlled by the sex chromosomes. Humans have 22 pairs of autosomes and one pair of sex chromosomes. All act in the same way during cell division. For information on sex-linked characteristics.

Chromosome breakage is the physical breakage of subunits of a chromosome. It is usually followed by reunion (frequently at a foreign site, resulting in a chromosome unlike the original). Breakage and reunion of homologous chromosomes during meiosis are the basis for the classical model of crossing over, which results in unexpected types of offspring of a mating.

GENOTYPE

The genotype is the part of the genetic makeup of a cell, and therefore of any individual, which determines one of its characteristics (phenotype). The term was coined by the Danish botanist, plant physiologist and geneticist Wilhelm Johannsen in 1903.

Genotype is one of three factors that determine phenotype, along with inherited epigenetic factors and non-inherited environmental factors. Not all organisms with the same genotype look or act the same way because appearance and behavior are modified by environmental and growing conditions. Likewise, not all organisms that look alike necessarily have the same genotype.

One's genotype differs subtly from one's genomic sequence, because it refers to how an individual *differs* or is specialized within a group of individuals or a species. So, typically, one refers to an individual's genotype with regard to a particular gene of interest and the combination of alleles the individual carries. Genotypes are often denoted with letters, for example *Bb*, where *B* stands for one allele and *b* for another.

Somatic mutations which are acquired rather than inherited, such as those in cancers, are not part of the individual's genotype. Hence, scientists and physicians sometimes talk about the genotype of a particular cancer, that is, of the disease as distinct from the diseased.

An example of a characteristic determined by a genotype is the petal color in a pea plant. The collection of all genetic possibilities for a single trait are called alleles; two alleles for petal color are purple and white.

Mendelian Inheritance

The distinction between genotype and phenotype is commonly experienced when studying family patterns for certain hereditary diseases or conditions, for example, hemophilia. Humans and most animals are diploid; thus there are two alleles for any given gene. These alleles can be the same (homozygous) or different (heterozygous), depending on the individual. With a dominant allele, the offspring is guaranteed to inherit the trait in question irrespective of the second allele.

In the case of an albino with a recessive allele (aa), the phenotype depends upon the other allele (Aa, aA, aa or AA). An affected person mating with a heterozygous individual (Aa or aA, also carrier) there is a 50-50 chance the offspring will be albino's phenotype. If a heterozygote mates with another heterozygote, there is 75% chance passing the gene on and only a 25% chance that the gene will be displayed. A homozygous dominant (AA) individual has a normal phenotype and no risk of abnormal offspring. A homozygous recessive individual has an abnormal phenotype and is guaranteed to pass the abnormal gene onto offspring.

In the case of hemophilia, it is sex-linked thus only carried on the X chromosome. Only females can be a carrier in which the abnormality is not displayed. This woman has a normal phenotype, but runs a 50-50 chance, with an unaffected partner, of passing her abnormal gene on to her offspring. If she mated with a man with haemophilia (another carrier) there would be a 75% chance of passing on the gene.

Determination

Genotyping is the process of elucidating the genotype of an individual with a biological assay. Also known as a *genotypic assay*, techniques include PCR, DNA fragment analysis, allele specific oligonucleotide (ASO) probes, DNA sequencing, and nucleic acid hybridization to DNA microarrays or beads. Several common genotyping techniques include restriction fragment length polymorphism (*RFLP*), terminal restriction fragment length polymorphism (*t-RFLP*), amplified fragment length polymorphism (*AFLP*), and multiplex ligation-dependent probe amplification (*MLPA*).

DNA fragment analysis can also be used to determine such disease causing genetics aberrations as microsatellite instability (*MSI*), *trisomy* or aneuploidy, and loss of heterozygosity (*LOH*). MSI and LOH in particular have been associated with cancer cell genotypes for colon, breast and cervical cancer.

The most common chromosomal aneuploidy is a trisomy of chromosome 21 which manifests itself as Down syndrome. Current technological limitations typically allow only a fraction of an individual's genotype to be determined efficiently.

PHENOTYPE

The phenotype of an organism is the composite of the organism's observable characteristics or traits, including its morphology or physical form and structure; its developmentalprocesses; its biochemical and physiological properties; its behavior, and the products of behavior, for example, a bird's nest. An organism's phenotype results from two basic factors: the expression of an organism's genetic code, or its genotype, and the influence of environmental factors, which may interact, further affecting phenotype. When two or more clearly different phenotypes exist in the same population of a species, the species is called polymorphic. A well-documented polymorphism is Labrador Retriever coloring; while the coat color depends on many genes, it is clearly seen in the environment as yellow, black and brown. Richard Dawkins in 1978 and then again in his 1982 book *The Extended Phenotype* suggested that bird nests and other built structures such as caddis fly larvae cases and beaver dams can be considered as "extended phenotypes".

The shells of individuals within the bivalve mollusk species *Donax variabilis*show diverse coloration and patterning in their phenotypes.

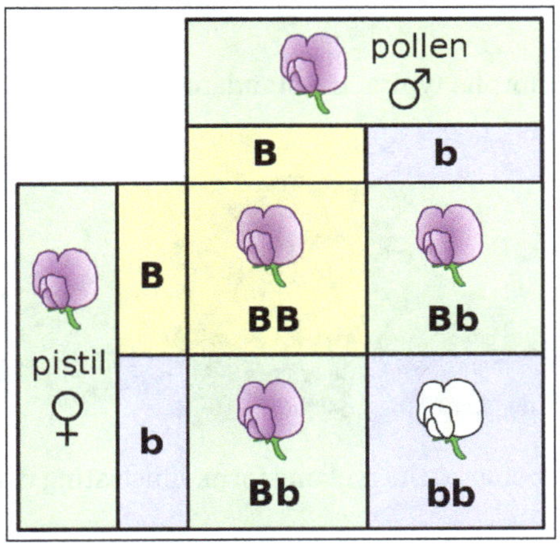

Here the relation between genotype and phenotype is illustrated, using a Punnett square, for the character of petal colour in a pea plant. The letters B and b represent alleles for colour and the pictures show the resultant flowers.

The genotype-phenotype distinction was proposed by Wilhelm Johannsen in 1911 to make clear the difference between an organism's heredity and what that heredity produces. The distinction is similar to that proposed by August Weismann, who distinguished between germ plasm (heredity) and somatic cells (the body).

The genotype-phenotype distinction should not be confused with Francis Crick's central dogma of molecular biology, which is a statement about the directionality of molecular sequential information flowing from DNA to protein, and not the reverse.

The term "phenotype" has sometimes been incorrectly used as a shorthand for phenotypic difference from wild type, bringing the absurd statement that a mutation has no phenotype. Despite its seemingly straightforward definition, the concept of the phenotype has hidden subtleties. It may seem that anything dependent on the genotype is a phenotype, including molecules such as RNA and proteins. Most molecules and structures coded by the genetic material are not visible in the appearance of an organism, yet they are observable (for example by Western blotting) and are thus part of the phenotype; human blood groupsare an example. It may seem that this goes beyond the original intentions of the concept with its focus on the (living) organism in itself. Either way, the term phenotype includes inherent traits or characteristics that are observable or traits that can be made visible by some technical procedure. A notable extension to this idea is the presence of "organic molecules" or metabolites that are generated by organisms from chemical reactions of enzymes.

Another extension adds behavior to the phenotype, since behaviors are observable characteristics. Behavioral phenotypes include cognitive, personality, and behavioral patterns. Some behavioral phenotypes may characterize psychiatric disorders or syndromes.

Biston betularia morpha typica, the standard light-colored peppered moth.

B.betularia morpha carbonaria, the melanic form, illustrating discontinuous variation.

Phenotypic Variation

Phenotypic variation (due to underlying heritable genetic variation) is a fundamental prerequisite for evolution by natural selection. It is the living organism as a whole that contributes (or not) to the next generation, so natural selection affects the genetic structure of a population indirectly via the contribution of phenotypes. Without phenotypic variation, there would be no evolution by natural selection.

The interaction between genotype and phenotype has often been conceptualized by the following relationship:

Genotype (G) + environment (E) → phenotype (P).

A more nuanced version of the relationship is:

Genotype (G) + environment (E) + genotype & environment interactions (GE) → phenotype (P).

Genotypes often have much flexibility in the modification and expression of phenotypes; in many organisms these phenotypes are very different under varying environmental conditions. The plant *Hieracium umbellatum* is found growing in two different habitats in Sweden. One habitat is rocky, sea-side cliffs, where the plants are bushy with broad leaves and expanded inflorescences; the other is among sand dunes where the plants grow prostrate with narrow leaves and compact inflorescences. These habitats alternate along the coast of Sweden and the habitat that the seeds of *Hieracium umbellatum* land in, determine the phenotype that grows.

An example of random variation in *Drosophila* flies is the number of ommatidia, which may vary (randomly) between left and right eyes in a single individual as much as they do between different genotypes overall, or between clones raised in different environments.

The concept of phenotype can be extended to variations below the level of the gene that affect an organism's fitness. For example, silent mutations that do not change the corresponding amino acid sequence of a gene may change the frequency of guanine-cytosine base pairs (GC content). These base pairs have a higher thermal stability (*melting point*) than adenine-thymine, a property that might convey, among organisms living in high-temperature environments, a selective advantage on variants enriched in GC content.

Extended Phenotype

Richard Dawkins described a phenotype that included all effects that a gene has on its surroundings, including other organisms, as an extended phenotype, arguing that "An animal's behavior tends to maximize the survival of the genes ‹for' that behavior, whether or not those genes happen to be in the body of the particular animal performing it." For instance, an organism such as a beaver modifies its environment by building a beaver dam; this can be considered an expression of its genes, just as its incisor teeth are—which it uses to modify its environment. Similarly, when a bird feeds a brood parasite such as a cuckoo, it is unwittingly extending its phenotype; and when genes in an orchid affect orchid bee behavior to increase pollination, or when genes in a peacock affect the copulatory decisions of peahens, again, the phenotype is being extended. Genes are, in Dawkins's view, selected by their phenotypic effects.

Other biologists broadly agree that the extended phenotype concept is relevant, but consider that its role is largely explanatory, rather than assisting in the design of experimental tests.

Phenome and Phenomics

Although a phenotype is the ensemble of observable characteristics displayed by an organism, the word *phenome* is sometimes used to refer to a collection of traits, while the simultaneous study of such a collection is referred to as *phenomics*. Phenomics is an important field of study because it can be used to figure out which genomic variants affect phenotypes which then can be used to explain things like health, disease, and evolutionary fitness. Phenomics forms a large part of the Human Genome Project.

Phenomics has widespread applications in the agricultural industry. With an exponentially growing population and inconsistent weather patterns due to global warming, it has become increasingly difficult to cultivate enough crops to support the world's population. Advantageous genomic variations, like drought and heat resistance, can be identified through the use of phenomics to create more durable GMOs.

Phenomics is also a crucial stepping stone towards personalized medicine, particularly drug therapy. This application of phenomics has the greatest potential to avoid testing drug therapies that will prove to be ineffective or unsafe. Once the phenomic database has acquired more data, patient phenomic information can be used to select specific drugs tailored to the patient. As the regulation of phenomics develops there is a potential that new knowledge bases will help achieve the promise of personalized medicine and treatment of neuropsychiatric syndromes.

GENETIC CARRIER

	A	a
A	AA	Aa
a	Aa	aa

Punnett square depicting a cross between two genetic carriers. The chance of two genetic carriers having a child with two copies of the recessive gene, thus being homozygous recessive, is 25%.

A hereditary carrier (or just carrier), is a person or other organism that has inherited a recessive allele for a genetic trait or mutation but usually does not display that trait or show symptoms of the disease. Carriers are, however, able to pass the allele onto their offspring, who may then express the genetic if they inherit the recessive allele from both parents. The chance of two carriers having a child with the disease is 25%. This phenomenon is a direct result of the recessive nature of many genes.

X-linked Genes

Queen Victoria, and her daughters Princesses Alice and Beatrix, were carriers of the X-linked hemophilia gene (an abnormal allele of a gene, necessary to produce one of the blood clotting factors). Both had children who continued to pass on the gene to succeeding generations of the royal houses of Spain and Russia, into which they married. Since males only have one X chromosome, males who carried the altered gene had hemophilia B. Females have two X chromosomes, so one copy of an X-linked recessive gene would cause them to be an asymptomatic carrier. These females simply passed it to half of their children.

Cystic Fibrosis

Up to 1 in 25 individuals of Northern European ancestry may be considered carriers of mutations (CFTR genes) that can lead to cystic fibrosis. The disease appears only when two of these carriers have children, as each pregnancy between them will have a 25% chance of producing a child with the disease. Although only about 1 of every 3,000 Caucasian newborns has CF, there are more than 900 known mutations of the gene that causes CF. Current tests look for the most common mutations.

Genetic testing can be used to tell if a person carries one or more mutations of the CF gene and how many copies of each mutation. The test looks at a person's DNA, which is taken from cells in a blood sample or from cells that are gently scraped from inside the mouth.

The mutations screened by the test vary according to a person's ethnic group or by the occurrence of CF already in the family. More than 10 million Americans, including 1 in 25 Caucasian Americans, are carriers of one mutation of the CF gene. CF is present in other races, though not as frequently as in Caucasian individuals. 1 in 46 Hispanic Americans, 1 in 65 African Americans, and 1 in 90 Asian Americans carry a mutation of the CF gene.

Sickle Cell Anemia

Sickle cell anemia is the most common genetic disorder among African Americans in the United States. While approximately 8% are carriers, 1 in 375 African Americans are born with the disease. Carriers are typically asymptomatic, but they may show symptoms at high altitudes or under oxygen-poor environments as in instances of extreme exercise. Carriers are also known to be resistant to malaria, suggesting there is a heterozygote advantage in certain regions of Africa. This is a probable explanation for why the disease is most prevalent among African Americans.

GENETIC CODE

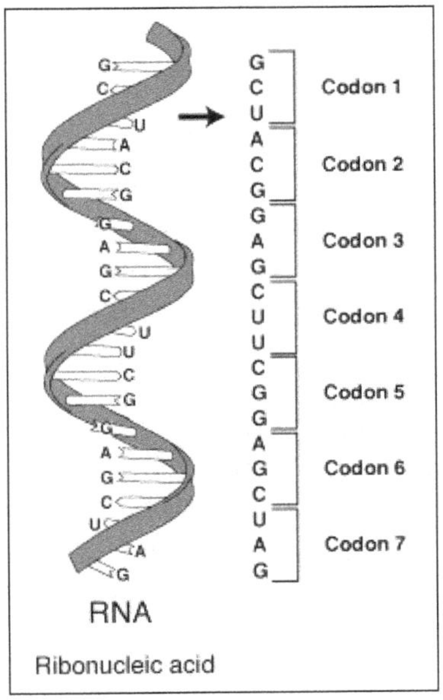

A series of codons in part of a messenger RNA (mRNA) molecule. Each codon consists of three nucleotides, usually corresponding to a single amino acid. The nucleotides are abbreviated with the letters A, U, G and C. This is mRNA, which uses U (uracil). DNA uses T (thymine) instead. This mRNA molecule will instruct a ribosome to synthesize a protein according to this code.

The genetic code is the set of rules used by living cells to translateinformation encoded within genetic material (DNA or mRNA sequences) into proteins. Translation is accomplished by the ribosome, which links amino acids in an order specified by messenger RNA (mRNA), using transfer RNA (tRNA) molecules to carry amino acids and to read the mRNA three nucleotides at a time. The genetic code is highly similar among all organisms and can be expressed in a simple table with 64 entries.

The code defines how sequences of nucleotide triplets, called *codons*, specify which amino acid will be added next during protein synthesis. With some exceptions, a three-nucleotide codon in a nucleic acid sequence specifies a single amino acid. The vast majority of genes are encoded with a single scheme. That scheme is often referred to as the canonical or standard genetic code, or simply *the* genetic code, though variant codes (such as in human mitochondria) exist. While the "genetic code" determines a protein's amino acid sequence, other genomic regions determine when and where these proteins are produced according to various "gene regulatory codes".

The genetic code is a key part of the story of life, according to which self-replicating RNA molecules preceded life as we know it. The main hypothesis for life's origin is the RNA world hypothesis. Any model for the emergence of genetic code is intimately related to a model of the transfer from ribozymes (RNA enzymes) to proteins as the principal enzymes in cells. In line with the RNA world hypothesis, transfer RNA molecules appear to have evolved before modern aminoacyl-tRNA synthetases, so the latter cannot be part of the explanation of its patterns.

A hypothetical randomly evolved genetic code further motivates a biochemical or evolutionary model for its origin. If amino acids were randomly assigned to triplet codons, there would be 1.5×10^{84} possible genetic codes. This number is found by calculating the number of ways that 21 items (20 amino acids plus one stop) can be placed in 64 bins, wherein each item is used at least once. However, the distribution of codon assignments in the genetic code is nonrandom. In particular, the genetic code clusters certain amino acid assignments.

Amino acids that share the same biosynthetic pathway tend to have the same first base in their codons. This could be an evolutionary relic of an early, simpler genetic code with fewer amino acids that later evolved to code a larger set of amino acids. It could also reflect steric and chemical properties that had another effect on the codon during its evolution. Amino acids with similar physical properties also tend to have similar codons,reducing the problems caused by point mutations and mistranslations.

Given the non-random genetic triplet coding scheme, a tenable hypothesis for the origin of genetic code could address multiple aspects of the codon table, such as absence of codons for D-amino acids, secondary codon patterns for some amino acids, confinement of synonymous positions to third position, the small set of only 20 amino acids (instead of a number approaching 64), and the relation of stop codon patterns to amino acid coding patterns.

Three main hypotheses address the origin of the genetic code. Many models belong to one of them or to a hybrid:

- Random freeze: The genetic code was randomly created. For example, early tRNA-like ribozymes may have had different affinities for amino acids, with codons emerging from another part of the ribozyme that exhibited random variability. Once enough peptides were coded for, any major random change in the genetic code would have been lethal; hence it became "frozen".

- Stereochemical affinity: The genetic code is a result of a high affinity between each amino acid and its codon or anti-codon; the latter option implies that pre-tRNA molecules matched their corresponding amino acids by this affinity. Later during evolution, this matching was gradually replaced with matching by aminoacyl-tRNA synthetases.
- Optimality: The genetic code continued to evolve after its initial creation, so that the current code maximizes some fitness function, usually some kind of error minimization.

Hypotheses have addressed a variety of scenarios:

- Chemical principles govern specific RNA interaction with amino acids: Experiments with aptamers showed that some amino acids have a selective chemical affinity for their codons. Experiments showed that of 8 amino acids tested, 6 show some RNA triplet-amino acid association.
- Biosynthetic expansion: The genetic code grew from a simpler earlier code through a process of "biosynthetic expansion". Primordial life "discovered" new amino acids (for example, as by-products of metabolism) and later incorporated some of these into the machinery of genetic coding. Although much circumstantial evidence has been found to suggest that fewer amino acid types were used in the past,precise and detailed hypotheses about which amino acids entered the code in what order are controversial.
- Natural selection has led to codon assignments of the genetic code that minimize the effects of mutations.A recent hypothesis suggests that the triplet code was derived from codes that used longer than triplet codons (such as quadruplet codons). Longer than triplet decoding would increase codon redundancy and would be more error resistant. This feature could allow accurate decoding absent complex translational machinery such as the ribosome, such as before cells began making ribosomes.
- Information channels: Information-theoretic approaches model the process of translating the genetic code into corresponding amino acids as an error-prone information channel. The inherent noise (that is, the error) in the channel poses the organism with a fundamental question: how can a genetic code be constructed to withstand noise while accurately and efficiently translating information? These "rate-distortion" models suggest that the genetic code originated as a result of the interplay of the three conflicting evolutionary forces: the needs for diverse amino acids, for error-tolerance and for minimal resource cost. The code emerges at a transition when the mapping of codons to amino acids becomes nonrandom. The code's emergence is governed by the topology defined by the probable errors and is related to the map coloring problem.
- Game theory: Models based on signaling games combine elements of game theory, natural selection and information channels. Such models have been used to suggest that the first polypeptides were likely short and had non-enzymatic function. Game theoretic models

suggested that the organization of RNA strings into cells may have been necessary to prevent "deceptive" use of the genetic code, i.e. preventing the ancient equivalent of viruses from overwhelming the RNA world.

- Stop codons: Codons for translational stops are also an interesting aspect to the problem of the origin of the genetic code. As an example for addressing stop codon evolution, it has been suggested that the stop codons are such that they are most likely to terminate translation early in the case of a frame shift error. In contrast, some stereochemical molecular models explain the origin of stop codons as "unassignable".

Features

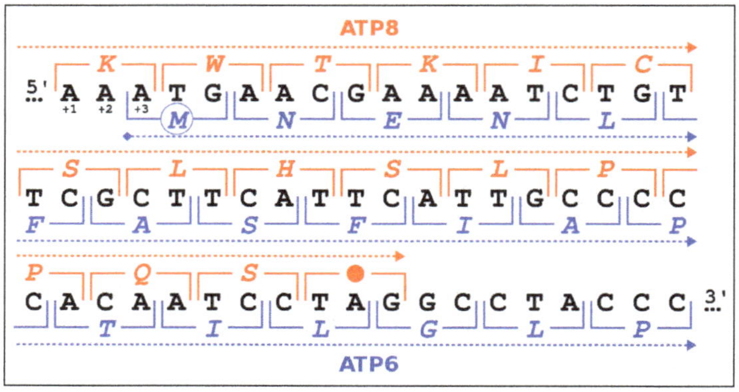

Reading frames in the DNA sequence of a region of the human mitochondrial genome coding for the genes *MT-ATP8* and *MT-ATP6* (in black: positions 8,525 to 8,580 in the sequence accession NC_012920). There are three possible reading frames in the 5' → 3' forward direction, starting on the first (+1), second (+2) and third position (+3). For each codon (square brackets), the amino acid is given by the vertebrate mitochondrial code, either in the +1 frame for *MT-ATP8* (in red) or in the +3 frame for *MT-ATP6* (in blue). The *MT-ATP8* genes terminates with the TAG stop codon (red dot) in the +1 frame. The *MT-ATP6*gene starts with the ATG codon (blue circle for the M amino acid) in the +3 frame.

Reading Frame

A reading frame is defined by the initial triplet of nucleotides from which translation starts. It sets the frame for a run of successive, non-overlapping codons, which is known as an "open reading frame" (ORF). For example, the string 5'-AAATGAACG-3', if read from the first position, contains the codons AAA, TGA, and ACG ; if read from the second position, it contains the codons AAT and GAA ; and if read from the third position, it contains the codons ATG and AAC. Every sequence can, thus, be read in its 5' → 3' direction in three reading frames, each producing a possibly distinct amino acid sequence: in the given example, Lys (K)-Trp (W)-Thr (T), Asn (N)-Glu (E), or Met (M)-Asn (N), respectively (when translating with the vertebrate mitochondrial code). When DNA is double-stranded, six possible reading frames are defined, three in the forward orientation on one strand and three reverse on the opposite strand. Protein-coding frames are defined by a start codon, usually the first AUG (ATG) codon in the RNA (DNA) sequence.

In eukaryotes, ORFs in exons are often interrupted by introns.

Start/Stop Codons

Translation starts with a chain-initiation codon or start codon. The start codon alone is not sufficient to begin the process. Nearby sequences such as the Shine-Dalgarno sequence in *E. coli* and initiation factors are also required to start translation. The most common start codon is AUG, which is read as methionine or, in bacteria, as formylmethionine. Alternative start codons depending on the organism include "GUG" or "UUG"; these codons normally represent valine and leucine, respectively, but as start codons they are translated as methionine or formylmethionine.

The three stop codons have names: UAG is *amber*, UGA is *opal* (sometimes also called *umber*), and UAA is *ochre.* Stop codons are also called "termination" or "nonsense" codons. They signal release of the nascent polypeptide from the ribosome because no cognate tRNA has anticodons complementary to these stop signals, allowing a release factor to bind to the ribosome instead.

Effect of Mutations

During the process of DNA replication, errors occasionally occur in the polymerization of the second strand. These errors, mutations, can affect an organism's phenotype, especially if they occur within the protein coding sequence of a gene. Error rates are typically 1 error in every 10–100 million bases—due to the "proofreading" ability of DNA polymerases.

Missense mutations and nonsense mutations are examples of point mutations that can cause genetic diseases such as sickle-cell disease and thalassemiarespectively. Clinically important missense mutations generally change the properties of the coded amino acid residue among basic, acidic, polar or non-polar states, whereas nonsense mutations result in a stop codon.

Mutations that disrupt the reading frame sequence by indels (insertions or deletions) of a non-multiple of 3 nucleotide bases are known as frameshift mutations. These mutations usually result in a completely different translation from the original, and likely cause a stop codon to be read, which truncates the protein. These mutations may impair the protein's function and are thus rare in *in vivo* protein-coding sequences. One reason inheritance of frameshift mutations is rare is that, if the protein being translated is essential for growth under the selective pressures the organism faces, absence of a functional protein may cause death before the organism becomes viable. Frameshift mutations may result in severe genetic diseases such as Tay–Sachs disease.

Although most mutations that change protein sequences are harmful or neutral, some mutations have benefits. These mutations may enable the mutant organism to withstand particular environmental stresses better than wild type organisms, or reproduce more quickly. In these cases a mutation will tend to become more common in a population through natural selection. Viruses that use RNA as their genetic material have rapid mutation rates, which can be an advantage, since these viruses thereby evolve rapidly, and thus evade the immune system defensive responses. In large populations of asexually reproducing organisms, for example, *E. coli*, multiple beneficial mutations may co-occur. This phenomenon is called clonal interference and causes competition among the mutations.

Degeneracy

Degeneracy is the redundancy of the genetic code. This term was given by Bernfield and

Nirenberg. The genetic code has redundancy but no ambiguity. For example, although codons GAA and GAG both specify glutamic acid (redundancy), neither specifies another amino acid (no ambiguity). The codons encoding one amino acid may differ in any of their three positions. For example, the amino acid leucine is specified by YUR or CUN (UUA, UUG, CUU, CUC, CUA, or CUG) codons (difference in the first or third position indicated using IUPAC notation), while the amino acid serine is specified by UCN or AGY (UCA, UCG, UCC, UCU, AGU, or AGC) codons (difference in the first, second, or third position). A practical consequence of redundancy is that errors in the third position of the triplet codon cause only a silent mutation or an error that would not affect the protein because the hydrophilicityor hydrophobicity is maintained by equivalent substitution of amino acids; for example, a codon of NUN (where N = any nucleotide) tends to code for hydrophobic amino acids. NCN yields amino acid residues that are small in size and moderate in hydropathy; NAN encodes average size hydrophilic residues. The genetic code is so well-structured for hydropathy that a mathematical analysis (Singular Value Decomposition) of 12 variables (4 nucleotides x 3 positions) yields a remarkable correlation (C = 0.95) for predicting the hydropathy of the encoded amino acid directly from the triplet nucleotide sequence, *without translation*. Note in the table, below, eight amino acids are not affected at all by mutations at the third position of the codon, whereas in the figure above, a mutation at the second position is likely to cause a radical change in the physicochemical properties of the encoded amino acid. Nevertheless, changes in the first position of the codons are more important than changes in the second position on a global scale. The reason may be that charge reversal (from a positive to a negative charge or vice versa) can only occur upon mutations in the first position, but never upon changes in the second position of a codon. Such charge reversal may have dramatic consequences for the structure or function of a protein. This aspect may have been largely underestimated by previous studies.

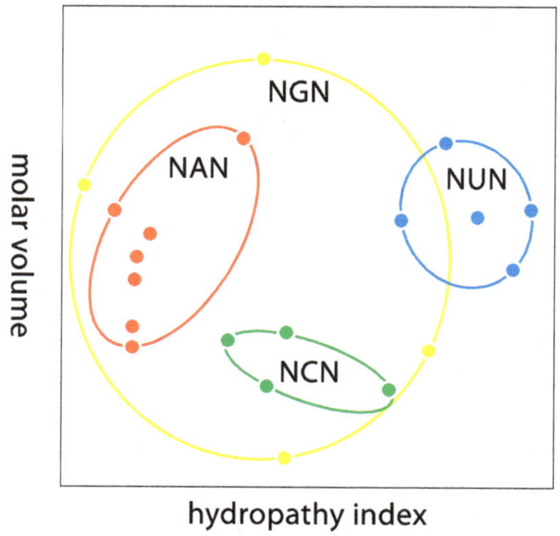

Grouping of codons by amino acid residue molar volume and hydropathy.

Codon usage Bias

The frequency of codons, also known as codon usage bias, can vary from species to species with functional implications for the control of translation. The following codon usage table is for the human genome.

Alternative Genetic Codes

Non-standard Amino Acids

In some proteins, non-standard amino acids are substituted for standard stop codons, depending on associated signal sequences in the messenger RNA. For example, UGA can code for selenocysteine and UAG can code for pyrrolysine. Selenocysteine became to be seen as the 21st amino acid, and pyrrolysine as the 22nd. Unlike selenocysteine, pyrrolysine-encoded UAG is translated with the participation of a dedicated aminoacyl-tRNA synthetase. Both selenocysteine and pyrrolysine may be present in the same organism. Although the genetic code is normally fixed in an organism, the achaeal prokaryote *Acetohalobium arabaticum* can expand its genetic code from 20 to 21 amino acids (by including pyrrolysine) under different conditions of growth.

Variations

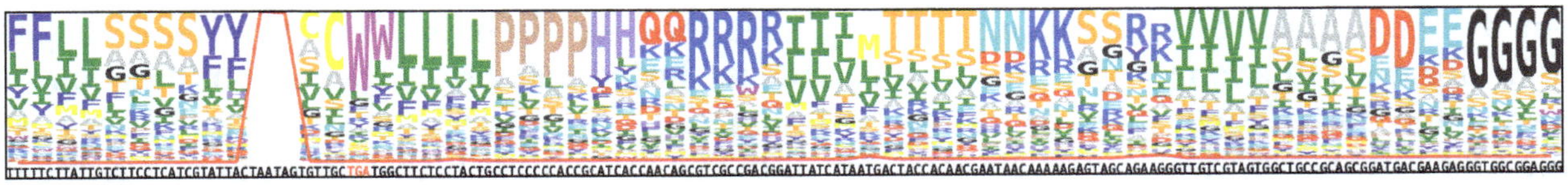

Genetic code logo of the Globobulimina pseudospinescens mitochondrial genome. The logo shows the 64 codons from left to right, predicted alternatives in red (relative to the standard genetic code). Red line: stop codons. The height of each amino acid in the stack shows how often it is aligned to the codon in homologous protein domains. The stack height indicates the support for the prediction.

Variations on the standard code were predicted in the 1970s. The first was discovered in 1979, by researchers studying human mitochondrial genes. Many slight variants were discovered thereafter, including various alternative mitochondrial codes. These minor variants for example involve translation of the codon UGA as tryptophan in *Mycoplasma* species, and translation of CUG as a serine rather than leucine in yeasts of the "CTG clade" (such as *Candida albicans*). Because viruses must use the same genetic code as their hosts, modifications to the standard genetic code could interfere with viral protein synthesis or functioning. However, viruses such as totiviruses have adapted to the host's genetic code modification. In bacteria and archaea, GUG and UUG are common start codons. In rare cases, certain proteins may use alternative start codons. Surprisingly, variations in the interpretation of the genetic code exist also in human nuclear-encoded genes: In 2016, researchers studying the translation of malate dehydrogenase found that in about 4% of the mRNAs encoding this enzyme the stop codon is naturally used to encode the amino acids tryptophan and arginine. This type of recoding is induced by a high-readthrough stop codon context and it is referred to as functional translational readthrough.

Variant genetic codes used by an organism can be inferred by identifying highly conserved genes encoded in that genome, and comparing its codon usage to the amino acids in homologous proteins of other organisms. For example, the program FACIL infers a genetic code by searching which amino acids in homologous protein domains are most often aligned to every codon. The resulting amino acid probabilities for each codon are displayed in a genetic code logo, that also shows the support for a stop codon.

Despite these differences, all known naturally occurring codes are very similar. The coding mechanism is the same for all organisms: three-base codons, tRNA, ribosomes, single direction reading and translating single codons into single amino acids.

MUTATION

Mutation is an alteration in the genetic material (the genome) of a cell of a living organism or of a virus that is more or less permanent and that can be transmitted to the cell's or the virus's descendants. (The genomes of organisms are all composed of DNA, whereas viral genomes can be of DNA or RNA; Mutation in the DNA of a body cell of a multicellular organism (somatic mutation) may be transmitted to descendant cells by DNA replication and hence result in a sector or patch of cells having abnormal function, an example being cancer. Mutations in egg or sperm cells (germinal mutations) may result in an individual offspring all of whose cells carry the mutation, which often confers some serious malfunction, as in the case of a human genetic disease such as cystic fibrosis. Mutations result either from accidents during the normal chemical transactions of DNA, often during replication, or from exposure to high-energy electromagnetic radiation (e.g., ultraviolet light or X-rays) or particle radiation or to highly reactive chemicals in the environment. Because mutations are random changes, they are expected to be mostly deleterious, but some may be beneficial in certain environments. In general, mutation is the main source of genetic variation, which is the raw material for evolution by natural selection.

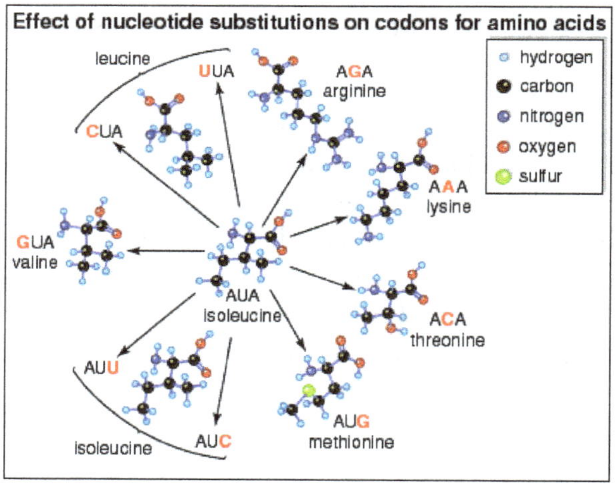

The effect of base substitutions, or point mutations, on the messenger-RNA codon AUA, which codes for the amino acid isoleucine. Substitutions (red letters) at the first, second, or third position in the codon can result in nine new codons corresponding to six different amino acids in addition to isoleucine itself. The chemical properties of some of these amino acids are quite different from those of isoleucine. Replacement of one amino acid in a protein by another can seriously affect the protein's biological function.

The genome is composed of one to several long molecules of DNA, and mutation can occur potentially anywhere on these molecules at any time. The most serious changes take place in the functional units of DNA, the genes. A mutated form of a gene is called a mutant allele. A gene is typically composed of a regulatory region, which is responsible for turning the gene's transcription on and off at the appropriate times during development, and a coding region, which carries the genetic code for the structure of a functional molecule, generally a protein. A protein is a chain of usually several hundred amino acids. Cells make 20 common amino acids, and it is the unique number and sequence of these that give a protein its specific function. Each amino acid is encoded by a unique sequence, or codon, of three of the four possible base pairs in the DNA (A–T, T–A, G–C, and C–G, the individual letters referring to the four nitrogenous bases adenine, thymine, guanine, and cytosine). Hence, a mutation that changes DNA sequence can change amino acid sequence and in this way potentially reduce or inactivate a protein's function. A change in the DNA sequence of a gene's regulatory region can adversely affect the timing and availability of the gene's protein and also lead to serious cellular malfunction. On the other hand, many mutations are silent, showing no obvious effect at the functional level. Some silent mutations are in the DNA between genes, or they are of a type that results in no significant amino acid changes.

Mutations are of several types. Changes within genes are called point mutations. The simplest kinds are changes to single base pairs, called base-pair substitutions. Many of these substitute an incorrect amino acid in the corresponding position in the encoded protein, and of these a large proportion result in altered protein function. Some base-pair substitutions produce a stop codon. Normally, when a stop codon occurs at the end of a gene, it stops protein synthesis, but, when it occurs in an abnormal position, it can result in a truncated and nonfunctional protein. Another type of simple change, the deletion or insertion of single base pairs, generally has a profound effect on the protein because the protein's synthesis, which is carried out by the reading of triplet codons in a linear fashion from one end of the gene to the other, is thrown off. This change leads to a frameshift in reading the gene such that all amino acids are incorrect from the mutation onward. More-complex combinations of base substitutions, insertions, and deletions can also be observed in some mutant genes.

Mutations that span more than one gene are called chromosomal mutations because they affect the structure, function, and inheritance of whole DNA molecules (microscopically visible in a coiled state as chromosomes). Often these chromosome mutations result from one or more coincident breaks in the DNA molecules of the genome (possibly from exposure to energetic radiation), followed in some cases by faulty rejoining. Some outcomes are large-scale deletions, duplications, inversions, and translocations. In a diploid species (a species, such as human beings, that has a double set of chromosomes in the nucleus of each cell), deletions and duplications alter gene balance and often result in abnormality. Inversions and translocations involve no loss or gain and are functionally normal unless a break occurs within a gene. However, at meiosis (the specialized nuclear divisions that take place during the production of gametes—i.e., eggs and sperm), faulty pairing of an inverted or translocated chromosome set with a normal set can result in gametes and hence progeny with duplications and deletions.

Loss or gain of whole chromosomes results in a condition called aneuploidy. One familiar result of aneuploidy is Down syndrome, a chromosomal disorder in which humans are born with an extra chromosome 21 (and hence bear three copies of that chromosome instead of the usual two). Another type of chromosome mutation is the gain or loss of whole chromosome sets. Gain of sets results in polyploidy—that is, the presence of three, four, or more chromosome sets instead of the usual two. Polyploidy has been a significant force in the evolution of new species of plants and animals.

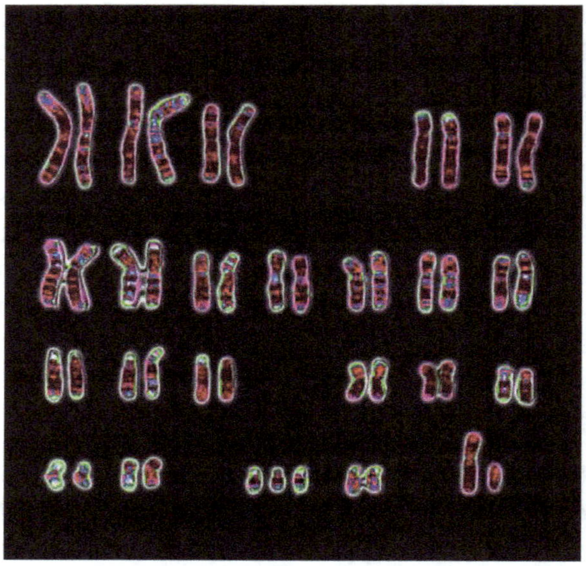

A karyotype of a human male with Down syndrome, showing a full chromosome complement plus an extra chromosome 21.

Most genomes contain mobile DNA elements that move from one location to another. The movement of these elements can cause mutation, either because the element arrives in some crucial location, such as within a gene, or because it promotes large-scale chromosome mutations via recombination between pairs of mobile elements in different locations.

At the level of whole populations of organisms, mutation can be viewed as a constantly dripping faucet introducing mutant alleles into the population, a concept described as mutational pressure. The rate of mutation differs for different genes and organisms. In RNA viruses, such as the human immunodeficiency virus, replication of the genome takes place within the host cell using a mechanism that is prone to error. Hence, mutation rates in such viruses are high. In general, however, the fate of individual mutant alleles is never certain. Most are eliminated by chance. In some cases a mutant allele can increase in frequency by chance, and then individuals expressing the allele can be subject to selection, either positive or negative. Hence, for any one gene the frequency of a mutant allele in a population is determined by a combination of mutational pressure, selection, and chance.

GENETIC VARIATION

Genetic variation can be described as the differences between organisms caused by alternate forms of DNA. Genetic variation in combination with environmental variation causes the total phenotypic variation seen in a population. The phenotypic variation is what is seen by the observer; the height of a plant for instance. The environmental variation is the difference in what each individual experiences. In our scenario, some plants may get more water and nutrients than others. If you remove the environmental differences from the overall variation, what you have left is the genetic variation. This variation represents the actual differences in DNA sequences between organisms.

Examples of Genetic Variation

Genetic Variation between Individuals

Look at the image of mussel shells below. All of these muscles belong to the same species, meaning they can all interbreed with each other. The differences in their patterns represents the total phenotypic variation in the population. Some of the variation comes from genetics, while some comes from the environment. To sort out what is genetic and what is environmental, scientist would have to conduct a series of experiments.

Coquina variation.

Two experiments would be required to find the overall genetic variation in the population. In the first, a single mussel would be cloned many times and placed in variable environments. The specimens would be allowed to grow and they would be observed in adulthood. Because their genetics are identical, the variation seen can be attributed solely to environmental variation. In the second experiment, the total variation in a population of wild mussels in the same environment must be observed. At the end of these two experiments, the scientist would have two numbers: one describing the environmental variance and one describing the phenotypic variance.

To get the genetic variation found in this population of mussels, the scientist would simply need to subtract the environmental variance observed with the clone from the total variance observed in the wild population. Another way of calculating the genetic variation is to sample the DNA of the population and measure the differences in the DNA directly. Since genetic variation is produced by differences in the DNA, these differences can be used in reverse to calculate the environmental variation in a population.

Genetic Variation between Species

While the above example discusses genetic variation between members of a population, the concept of genetic variation can be applied on a much grander scale. Consider for instance the Homeobox gene family. This family, known as the "Hox genes" for short, direct and coordinate the positions of body parts during development. These genes, or a variation of them is found among all bilaterally symmetrical animals. This includes everything from insects to fish and mammals. Scientist theorize that an early ancestor developed the Hox genes, which were quickly adapted to many forms of organism. The genetic variation represented in these genes is huge. They produce the different body types of most of the organism on Earth. However, they are still all related and the variance between them can be measured.

Sources of Genetic Variation

With all of the natural variety in the world, it is weird to think that all of the genetic variation comes from only a few simple sources. The simplest source is mutation. As DNA is exposed to the various chemicals and electromagnetic energies of the world, it can mutate. DNA is made of a specific sequence of nucleotides, which produce proteins. Mutations change these proteins by changing the sequence of nucleotides. While mutations are often thought of in a negative context, they drive evolution by putting forth new variants to be tested by the environment.

In fact, genetic variation is so important for species that many species reproduce sexually to aid the process of producing new varieties. Sexually reproducing organisms carry two copies of the genome, allowing mutations to lie dormant or express themselves more subtly. During sexual reproduction, genes are recombined in new ways. This process, known as recombination, shuffles the alleles present and allows different combinations to be expressed. This adds to the total genetic variation. When observing an isolated population, immigration can also be a source of genetic variation. Organism may bring new alleles that have established elsewhere and introduce them to the population.

Human Genetic Variation

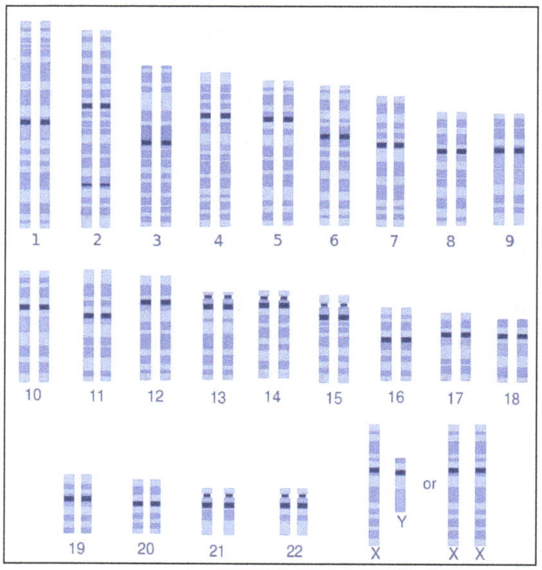

A graphical representation of the typical human karyotype.

Human genetic variation is the genetic differences in and among populations. There may be multiple variants of any given gene in the human population (alleles), a situation called polymorphism.

No two humans are genetically identical. Even monozygotic twins (who develop from one zygote) have infrequent genetic differences due to mutations occurring during development and gene copy-number variation. Differences between individuals, even closely related individuals, are the key to techniques such as genetic fingerprinting. As of 2017, there are a total of 324 million known variants from sequenced human genomes. As of 2015, the typical difference between the genomes of two individuals was estimated at 20 million base pairs (or 0.6% of the total of 3.2 billion base pairs).

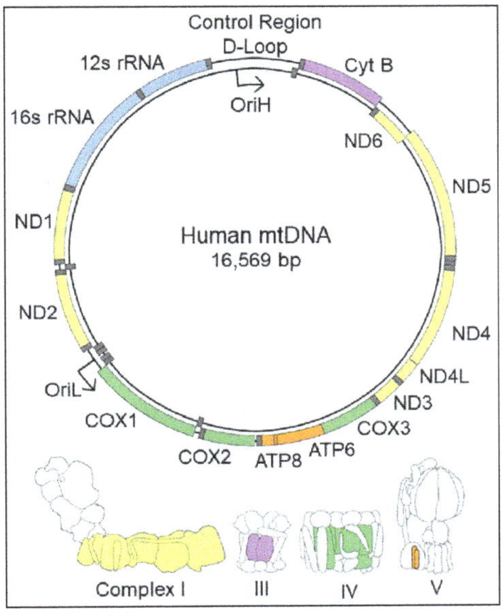

The human mitochondrial DNA.

Alleles occur at different frequencies in different human populations. Populations that are more geographically and ancestrally remote tend to differ more. The differences between populations represent a small proportion of overall human genetic variation. Populations also differ in the quantity of variation among their members. The greatest divergence between populations is found in sub-Saharan Africa, consistent with the recent African origin of non-African populations. Populations also vary in the proportion and locus of introgressed genes they received by archaic admixture both inside and outside of Africa.

The study of human genetic variation has evolutionary significance and medical applications. It can help scientists understand ancient human population migrations as well as how human groups are biologically related to one another. For medicine, study of human genetic variation may be important because some disease-causing alleles occur more often in people from specific geographic regions. New findings show that each human has on average 60 new mutations compared to their parents.

Causes of Variation

Causes of differences between individuals include independent assortment, the exchange of genes (crossing over and recombination) during reproduction (through meiosis) and various mutational events.

There are at least three reasons why genetic variation exists between populations. Natural selection may confer an adaptive advantage to individuals in a specific environment if an allele provides a competitive advantage. Alleles under selection are likely to occur only in those geographic regions where they confer an advantage. A second important process is genetic drift, which is the effect of random changes in the gene pool, under conditions where most mutations are neutral (that is, they do not appear to have any positive or negative selective effect on the organism). Finally, small migrant populations have statistical differences—call the founder effect—from the overall populations where they originated; when these migrants settle new areas, their descendant population typically differs from their population of origin: different genes predominate and it is less genetically diverse.

In humans, the main cause is genetic drift. Serial founder effects and past small population size (increasing the likelihood of genetic drift) may have had an important influence in neutral differences between populations. The second main cause of genetic variation is due to the high degree of neutrality of most mutations. A small, but significant number of genes appear to have undergone recent natural selection, and these selective pressures are sometimes specific to one region.

Measures of Variation

Genetic variation among humans occurs on many scales, from gross alterations in the human karyotype to single nucleotide changes. Chromosome abnormalities are detected in 1 of 160 live human births. Apart from sex chromosome disorders, most cases of aneuploidy result in death of the developing fetus (miscarriage); the most common extra autosomal chromosomes among live births are 21, 18 and 13.

Nucleotide diversity is the average proportion of nucleotides that differ between two individuals. As of 2004, the human nucleotide diversity was estimated to be 0.1% to 0.4% of base pairs. In 2015, the 1000 Genomes Project, which sequenced one thousand individuals from 26 human populations, found that "a typical [individual] genome differs from the reference human genome at 4.1 million to 5.0 million sites affecting 20 million bases of sequence"; the latter figure corresponds to 0.6% of total number of base pairs. Nearly all (>99.9%) of these sites are small differences, either single nucleotide polymorphisms or brief insertions or deletions (indels) in the genetic sequence, but structural variations account for a greater number of base-pairs than the SNPs and indels.

As of 2017, the Single Nucleotide Polymorphism Database (dbSNP), which lists SNP and other variants, listed 324 million variants found in sequenced human genomes.

Single Nucleotide Polymorphisms

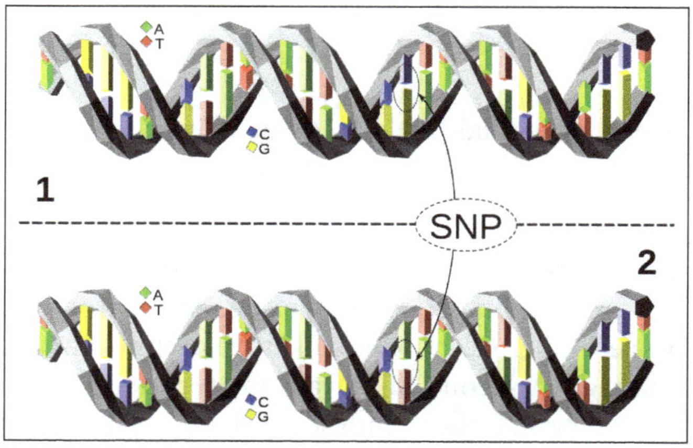

DNA molecule 1 differs from DNA molecule 2 at a single base-pair location (a C/T polymorphism).

A single nucleotide polymorphism (SNP) is a difference in a single nucleotide between members of one species that occurs in at least 1% of the population. The 2,504 individuals characterized by the 1000 Genomes Project had 84.7 million SNPs among them. SNPs are the most common type of sequence variation, estimated in 1998 to account for 90% of all sequence variants. Other sequence variations are single base exchanges, deletions and insertions. SNPs occur on average about every 100 to 300 bases and so are the major source of heterogeneity.

A functional, or non-synonymous, SNP is one that affects some factor such as gene splicing or messenger RNA, and so causes a phenotypic difference between members of the species. About 3% to 5% of human SNPs are functional. Neutral, or synonymous SNPs are still useful as genetic markers in genome-wide association studies, because of their sheer number and the stable inheritance over generations.

A coding SNP is one that occurs inside a gene. There are 105 Human Reference SNPs that result in premature stop codons in 103 genes. This corresponds to 0.5% of coding SNPs. They occur due to segmental duplication in the genome. These SNPs result in loss of protein, yet all these SNP alleles are common and are not purified in negative selection.

Structural Variation

Structural variation is the variation in structure of an organism's chromosome. Structural variations, such as copy-number variation and deletions, inversions, insertions and duplications, account for much more human genetic variation than single nucleotide diversity. This was concluded in 2007 from analysis of the diploid full sequences of the genomes of two humans: Craig Venter and James D. Watson. This added to the two haploidsequences which were amalgamations of sequences from many individuals, published by the Human Genome Project and Celera Genomics respectively.

According to the 1000 Genomes Project, a typical human has 2,100 to 2,500 structural variations, which include approximately 1,000 large deletions, 160 copy-number variants, 915 Alu insertions, 128 L1 insertions, 51 SVA insertions, 4 NUMTs, and 10 inversions.

Copy Number Variation

A copy-number variation (CNV) is a difference in the genome due to deleting or duplicating large regions of DNA on some chromosome. It is estimated that 0.4% of the genomes of unrelated humans differ with respect to copy number. When copy number variation is included, human-to-human genetic variation is estimated to be at least 0.5% (99.5% similarity). Copy number variations are inherited but can also arise during development.

A visual map with the regions with high genomic variation of the modern-human reference assembly relatively to a Neanderthal of 50k has been built by Pratas et al.

Epigenetics

Epigenetic variation is variation in the chemical tags that attach to DNA and affect how genes get read. The tags, "called epigenetic markings, act as switches that control how genes can be read." At some alleles, the epigenetic state of the DNA, and associated phenotype, can be inherited across generations of individuals.

Genetic Variability

Genetic variability is a measure of the tendency of individual genotypes in a population to vary (become different) from one another. Variability is different from genetic diversity, which is the amount of variation seen in a particular population. The variability of a trait is how much that trait tends to vary in response to environmental and genetic influences.

Clines

In biology, a cline is a continuum of species, populations, races, varieties, or forms of organisms that exhibit gradual phenotypic and/or genetic differences over a geographical area, typically as a result of environmental heterogeneity. In the scientific study of human genetic variation, a gene cline can be rigorously defined and subjected to quantitative metrics.

Haplogroups

In the study of molecular evolution, a haplogroup is a group of similar haplotypes that share a common ancestorwith a single nucleotide polymorphism (SNP) mutation. Haplogroups pertain to deep ancestral origins dating back thousands of years.

The most commonly studied human haplogroups are Y-chromosome (Y-DNA) haplogroups and mitochondrial DNA (mtDNA) haplogroups, both of which can be used to define genetic populations. Y-DNA is passed solely along the patrilineal line, from father to son, while mtDNA is passed down the matrilineal line, from mother to both daughter or son. The Y-DNA and mtDNA may change by chance mutation at each generation.

Variable Number Tandem Repeats

A variable number tandem repeat (VNTR) is the variation of length of a tandem repeat. A tandem repeat is the adjacent repetition of a short nucleotide sequence. Tandem repeats exist on many chromosomes, and their length varies between individuals. Each variant acts as an inherited allele, so they are used for personal or parental identification. Their analysis is useful in genetics and biology research, forensics, and DNA fingerprinting.

Short tandem repeats (about 5 base pairs) are called microsatellites, while longer ones are called minisatellites.

Categorization of the World Population

New data on human genetic variation has reignited the debate about a possible biological basis for categorization of humans into races. Most of the controversy surrounds the question of how to interpret the genetic data and whether conclusions based on it are sound. Some researchers argue that self-identified race can be used as an indicator of geographic ancestry for certain health risks and medications.

Although the genetic differences among human groups are relatively small, these differences in certain genes such as duffy, ABCC11, SLC24A5, called ancestry-informative markers (AIMs) nevertheless can be used to reliably situate many individuals within broad, geographically based groupings. For example, computer analyses of hundreds of polymorphic loci sampled in globally distributed populations have revealed the existence of genetic clustering that roughly is associated with groups that historically have occupied large continental and subcontinental regions.

Some commentators have argued that these patterns of variation provide a biological justification for the use of traditional racial categories. They argue that the continental clusterings correspond roughly with the division of human beings into sub-Saharan Africans; Europeans, Western Asians, Central Asians, Southern Asians and Northern Africans; Eastern Asians, Southeast Asians, Polynesians and Native Americans; and other inhabitants of Oceania. Other observers disagree, saying that the same data undercut traditional notions of racial groups. They point out, for example, that major populations considered races or subgroups within races do not necessarily form their own clusters.

Furthermore, because human genetic variation is clinal, many individuals affiliate with two or more continental groups. Thus, the genetically based "biogeographical ancestry" assigned to any given person generally will be broadly distributed and will be accompanied by sizable uncertainties.

In many parts of the world, groups have mixed in such a way that many individuals have relatively recent ancestors from widely separated regions. Although genetic analyses of large numbers of loci can produce estimates of the percentage of a person's ancestors coming from various continental populations, these estimates may assume a false distinctiveness of the parental populations, since human groups have exchanged mates from local to continental scales throughout history. Even with large numbers of markers, information for estimating admixture proportions of individuals or groups is limited, and estimates typically will have wide confidence intervals.

Genetic Clustering

Genetic data can be used to infer population structure and assign individuals to groups that often correspond with their self-identified geographical ancestry. Jorde and Wooding argued that "Analysis of many loci now yields reasonably accurate estimates of genetic similarity among individuals, rather than populations. Clustering of individuals is correlated with geographic origin or ancestry." However, identification by geographic origin may quickly break down when considering historical ancestry shared between individuals back in time.

An analysis of autosomal SNP data from the International HapMap Project (Phase II) and CEPH Human Genome Diversity Panel samples was published in 2009. The study of 53 populations taken from the HapMap and CEPH data (1138 unrelated individuals) suggested that natural selection may shape the human genome much more slowly than previously thought, with factors such as migration within and among continents more heavily influencing the distribution of genetic variations. A similar study published in 2010 found strong genome-wide evidence for selection due to changes in ecoregion, diet, and subsistence particularly in connection with polar ecoregions, with foraging, and with a diet rich in roots and tubers. In a 2016 study, principal component analysis of genome-wide data was capable of recovering previously-known targets for positive selection (without prior definition of populations) as well as a number of new candidate genes.

Forensic Anthropology

Forensic anthropologists can determine aspects of geographic ancestry (i.e. Asian, African, or European) from skeletal remains with a high degree of accuracy by analyzing skeletal measurements. According to some studies, individual test methods such as mid-facial measurements and femur traits can identify the geographic ancestry and by extension the racial category to which an individual would have been assigned during their lifetime, with over 80% accuracy, and in combination can be even more accurate. However, the skeletons of people who have recent ancestry in different geographical regions can exhibit characteristics of more than one ancestral group and, hence, cannot be identified as belonging to any single ancestral group.

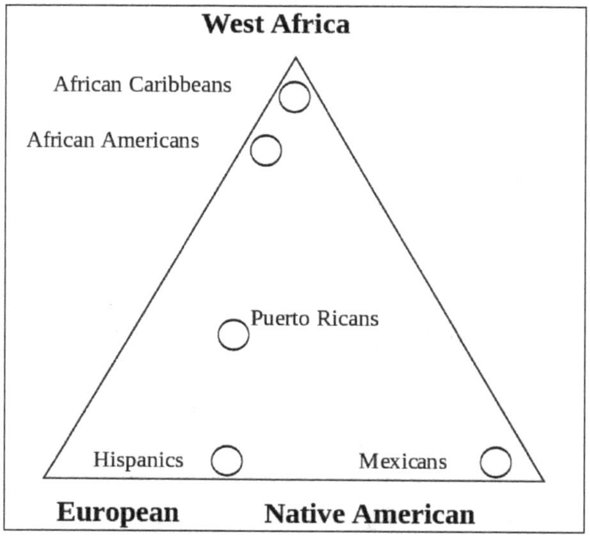

Triangle plot shows average admixture of five North American ethnic groups. Individuals that self-identify with each group can be found at many locations on the map, but on average groups tend to cluster differently.

Gene Flow and Admixture

Gene flow between two populations reduces the average genetic distance between the populations, only totally isolated human populations experience no gene flow and most populations have continuous gene flow with other neighboring populations which create the clinal distribution observed for moth genetic variation. When gene flow takes place between well-differentiated genetic populations the result is referred to as "genetic admixture".

Admixture mapping is a technique used to study how genetic variants cause differences in disease rates between population. Recent admixture populations that trace their ancestry to multiple continents are well suited for identifying genes for traits and diseases that differ in prevalence between parental populations. African-American populations have been the focus of numerous population genetic and admixture mapping studies, including studies of complex genetic traits such as white cell count, body-mass index, prostate cancer and renal disease.

An analysis of phenotypic and genetic variation including skin color and socio-economic status was carried out in the population of Cape Verde which has a well documented history of contact between Europeans and Africans, which showed that pattern of admixture in this population has been sex-biased and there is a significant interactions between socio economic status and skin color independent of the skin color and ancestry. Another study shows an increased risk of graft-versus-host disease complications after transplantation due to genetic variants in human leukocyte antigen (HLA) and non-HLA proteins.

GENE EXPRESSION

Genes encode proteins and proteins dictate cell function. Therefore, the thousands of genes expressed in a particular cell determine what that cell can do. Moreover, each step in the flow of information from DNA to RNA to protein provides the cell with a potential control point for self-regulating its functions by adjusting the amount and type of proteins it manufactures.

At any given time, the amount of a particular protein in a cell reflects the balance between that protein's synthetic and degradative biochemical pathways. On the synthetic side of this balance, recall that protein production starts at transcription (DNA to RNA) and continues with translation (RNA to protein). Thus, control of these processes plays a critical role in determining what proteins are present in a cell and in what amounts. In addition, the way in which a cell processes its RNA transcripts and newly made proteins also greatly influences protein levels.

How is Gene Expression Regulated?

The amounts and types of mRNA molecules in a cell reflect the function of that cell. In fact, thousands of transcripts are produced every second in every cell. Given this statistic, it is not surprising that the primary control point for gene expression is usually at the very beginning of the protein production process — the initiation of transcription. RNA transcription makes an efficient control point because many proteins can be made from a single mRNA molecule.

Transcript processing provides an additional level of regulation for eukaryotes, and the presence of a nucleus makes this possible. In prokaryotes, translation of a transcript begins before the transcript is complete, due to the proximity of ribosomes to the new mRNA molecules. In eukaryotes, however, transcripts are modified in the nucleus before they are exported to the cytoplasm for translation.

Eukaryotic transcripts are also more complex than prokaryotic transcripts. For instance, the primary transcripts synthesized by RNA polymerase contain sequences that will not be part of the mature RNA. These intervening sequences are called introns, and they are removed before the mature mRNA leaves the nucleus. The remaining regions of the transcript, which include the protein-coding regions, are called exons, and they are spliced together to produce the mature mRNA. Eukaryotic transcripts are also modified at their ends, which affects their stability and translation.

Of course, there are many cases in which cells must respond quickly to changing environmental conditions. In these situations, the regulatory control point may come well after transcription. For example, early development in most animals relies on translational control because very little transcription occurs during the first few cell divisions after fertilization. Eggs therefore contain many maternally originated mRNA transcripts as a ready reserve for translation after fertilization.

On the degradative side of the balance, cells can rapidly adjust their protein levels through the enzymatic breakdown of RNA transcripts and existing protein molecules. Both of these actions result in decreased amounts of certain proteins. Often, this breakdown is linked to specific events in the cell. The eukaryotic cell cycle provides a good example of how protein breakdown is linked to cellular events. This cycle is divided into several phases, each of which is characterized by distinct cyclin proteins that act as key regulators for that phase. Before a cell can progress from one phase of the cell cycle to the next, it must degrade the cyclin that characterizes that particular phase of the cycle. Failure to degrade a cyclin stops the cycle from continuing.

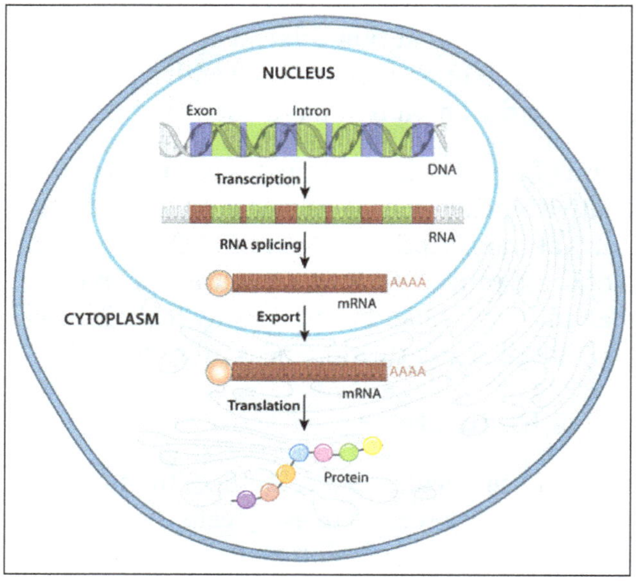

A schematic of a eukaryotic cell and its interior shows the transcription of DNA to RNA, and the translation of RNA to protein in four steps: transcription, RNA splicing, nuclear export, and translation. Each step is represented by a labeled arrow. Transcription of a DNA template to a pre-mRNA and the splicing of the pre-mRNA into a mature mRNA are shown inside the cell nucleus. The nuclear export brings the mature mRNA to the cytoplasm, where the mature mRNA message is translated into a protein.

How do Different Cells Express the Genes they Need?

Only a fraction of the genes in a cell are expressed at any one time. The variety of gene expression profiles characteristic of different cell types arise because these cells have distinct sets of transcription regulators. Some of these regulators work to increase transcription, whereas others prevent or suppress it.

Normally, transcription begins when an RNA polymerase binds to a so-called promoter sequence on the DNA molecule. This sequence is almost always located just upstream from the starting point for transcription (the 5' end of the DNA), though it can be located downstream of the mRNA (3' end). In recent years, researchers have discovered that other DNA sequences, known as enhancer sequences, also play an important part in transcription by providing binding sites for regulatory proteins that affect RNA polymerase activity. Binding of regulatory proteins to an enhancer sequence causes a shift in chromatin structure that either promotes or inhibits RNA polymerase and transcription factor binding. A more open chromatin structure is associated with active gene transcription. In contrast, a more compact chromatin structure is associated with transcriptional inactivity.

Some regulatory proteins affect the transcription of multiple genes. This occurs because multiple copies of the regulatory protein binding sites exist within the genome of a cell. Consequently, regulatory proteins can have different roles for different genes, and this is one mechanism by which cells can coordinate the regulation of many genes at once.

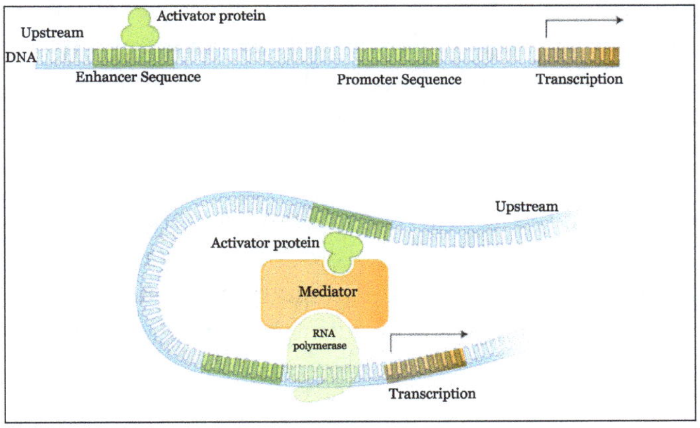

Modulation of transcription.

An activator protein bound to DNA at an upstream enhancer sequence can attract proteins to the promoter region that activate RNA polymerase (green) and thus transcription. The DNA can loop around on itself to cause this interaction between an activator protein and other proteins that mediate the activity of RNA polymerase.

How is Gene Expression Increased or Decreased in Response to Environmental Change?

In prokaryotes, regulatory proteins are often controlled by nutrient availability. This allows organisms such as bacteria to rapidly adjust their transcription patterns in response to environmental conditions. In addition, regulatory sites on prokaryotic DNA are typically located close to transcription promoter sites — and this plays an important part in gene expression.

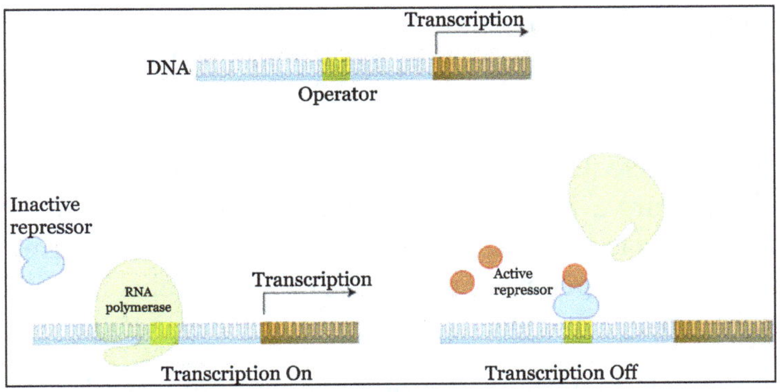

Transcription repression near the promoter region.

Molecules can interfere with RNA polymerase binding. An inactive repressor protein (blue) can become activated by another molecule (red circle). This active repressor can bind to a region near the promoter called an operator (yellow) and thus interfere with RNA polymerase binding to the promoter, effectively preventing transcription.

For an example of how this works, imagine a bacterium with a surplus of amino acids that signal the turning "on" of some genes and the turning "off" of others. In this particular example, cells might want to turn "on" genes for proteins that metabolize amino acids and turn "off" genes for proteins that synthesize amino acids. Some of these amino acids would bind to positive regulatory proteins called activators. Activator proteins bind to regulatory sites on DNA nearby to promoter regions that act as on/off switches. This binding facilitates RNA polymerase activity and transcription of nearby genes. At the same time, however, other amino acids would bind to negative regulatory proteins called repressors, which in turn bind to regulatory sites in the DNA that effectively block RNA polymerase binding.

The control of gene expression in eukaryotes is more complex than that in prokaryotes. In general, a greater number of regulatory proteins are involved, and regulatory binding sites may be located quite far from transcription promoter sites. Also, eukaryotic gene expression is usually regulated by a combination of several regulatory proteins acting together, which allows for greater flexibility in the control of gene expression.

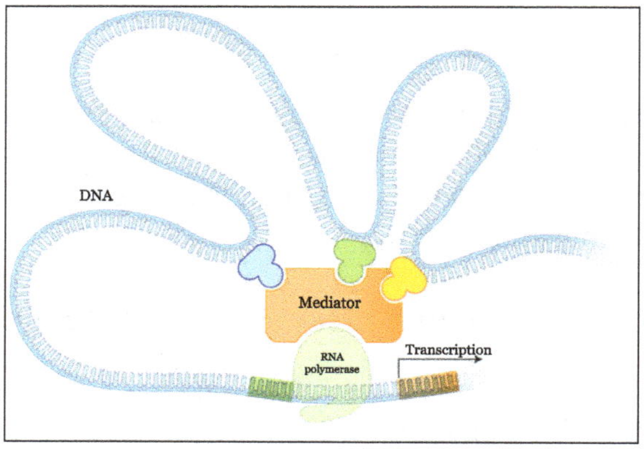

The complexity of multiple regulators.

Transcriptional regulators can each have a different role. Combinations of one, two, or three regulators (blue, green, and yellow shapes) can affect transcription in different ways by differentially affecting a mediator complex (orange), which is also composed of proteins. The effect is that the same gene can be transcribed in multiple ways, depending on the combination, presence, or absence of various transcriptional regulator proteins.

Enhancer sequences are DNA sequences that are bound by an activator protein and they can be located thousands of base pairs away from a promoter, either upstream or downstream from a gene. Activator protein binding is thought to cause DNA to loop out, bringing the activator protein into physical proximity with RNA polymerase and the other proteins in the complex that promote the initiation of transcription.

Different cell types express characteristic sets of transcriptional regulators. In fact, as multicellular organisms develop, different sets of cells within these organisms turn specific combinations of regulators on and off. Such developmental patterns are responsible for the variety of cell types present in the mature organism.

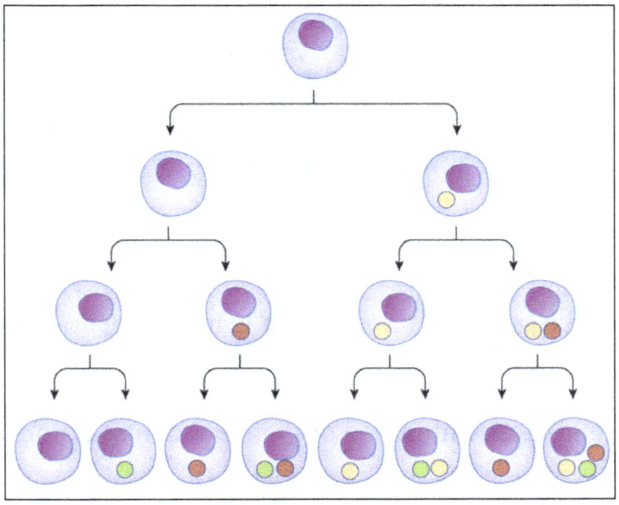

Transcriptional regulators can determine cell types.

The wide variety of cell types in a single organism can depend on different transcription factor activity in each cell type. Different transcription factors can turn on at different times during successive generations of cells. As cells mature and go through different stages (arrows), transcription factors (colored balls) can act on gene expression and change the cell in different ways. This change affects the next generation of cells derived from that cell. In subsequent generations, it is the combination of different transcription factors that can ultimately determine cell type.

References

- Genome-43, definition, scitable: nature.com, Retrieved 12 July, 2019
- "allele noun - Definition, pictures, pronunciation and usage notes". Oxford Advanced Learner's Dictionary. Retrieved 29 October 2017
- Nucleic-acid, science: britannica.com, Retrieved 2 June, 2019
- Allaby, Michael, ed. (2009). A dictionary of zoology (3rd ed.). Oxford: Oxford University Press. ISBN 9780199233410. OCLC 260204631
- Whats-a-genome, resources: genomenewsnetwork.org, Retrieved 1 August, 2019
- O'Brien, Gregory, ed. (2002). Behavioural Phenotypes in Clinical Practice. London: Mac Keith Press. ISBN 978-1-898683-27-8. Retrieved 27 September 2010
- Chromosome, science: britannica.com, Retrieved 3 May, 2019
- Edwards, Q. T.; Seibert, D.; Macri, C.; Carolyn, C.; Tilghman, J. (November 2004). "Assessing ethnicity in preconception counseling: Genetics--what nurse practitioners need to know". Clinical Practice. 16 (11): 472–480. Doi:10.1111/j.1745-7599.2004.tb00426.x.
- Whole-genome-sequencing, science: britannica.com, Retrieved 5 March, 2019
- Shu, Jian-Jun (2017). "A new integrated symmetrical table for genetic codes". Biosystems. 151: 21–26. Arxiv:1703.03787. Doi:10.1016/j.biosystems.2016.11.004. PMID 27887904
- Mutation-genetics, science: britannica.com, Retrieved 22 May, 2019
- "We are all mutants: First direct whole-genome measure of human mutation predicts 60 new mutations in each of us". Science Daily. 13 June 2011. Retrieved 5 September 2011

Genetic Modification and Engineering

CHAPTER 4

The direct manipulation of the genes of an organism using biotechnology is referred to as genetic engineering. It comprises of a group of technologies to change the genetic construction of cells. The topics elaborated in this chapter will help in gaining a better perspective about genetic engineering as well as various genetically modified organisms such as animals and crops.

Genetic engineering is the artificial the artificial manipulation, modification, and recombination of DNA or other nucleic acid molecules in order to modify an organism or population of organisms.

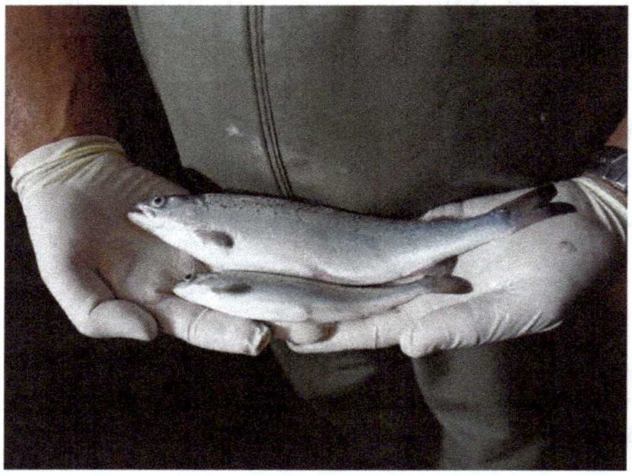

Genetically engineered salmon (top) and a natural salmon of the same age (bottom). The ability to engineer and precisely edit the genomes of animals, while potentially beneficial, has raised ethical questions.

The term genetic engineering initially referred to various techniques used for the modification or manipulation of organisms through the processes of heredity and reproduction. As such, the term embraced both artificial selection and all the interventions of biomedical techniques, among them artificial insemination, in vitro fertilization (e.g., "test-tube" babies), cloning, and gene manipulation. In the latter part of the 20th century, however, the term came to refer more specifically to methods of recombinant DNA technology (or gene cloning), in which DNA molecules from two or more sources are combined either within cells or in vitro and are then inserted into host organisms in which they are able to propagate.

The possibility for recombinant DNA technology emerged with the discovery of restriction enzymes in 1968 by Swiss microbiologist Werner Arber. The following year American microbiologist Hamilton O. Smith purified so-called type II restriction enzymes, which were found to be essential to genetic engineering for their ability to cleave a specific site within the DNA (as opposed to type I restriction enzymes, which cleave DNA at random sites). Drawing on Smith's work, American molecular biologist Daniel Nathans helped advance the technique of DNA recombination in 1970–71 and demonstrated that type II enzymes could be useful in genetic studies. Genetic engineering based on recombination was pioneered in 1973 by American biochemists Stanley N. Cohen and Herbert W. Boyer, who were among the first to cut DNA into fragments, rejoin different fragments, and insert the new genes into E. coli bacteria, which then reproduced.

Process and Techniques

Most recombinant DNA technology involves the insertion of foreign genes into the plasmids of common laboratory strains of bacteria. Plasmids are small rings of DNA; they are not part of the bacterium's chromosome (the main repository of the organism's genetic information). Nonetheless, they are capable of directing protein synthesis, and, like chromosomal DNA, they are reproduced and passed on to the bacterium's progeny. Thus, by incorporating foreign DNA (for example, a mammalian gene) into a bacterium, researchers can obtain an almost limitless number of copies of the inserted gene. Furthermore, if the inserted gene is operative (i.e., if it directs protein synthesis), the modified bacterium will produce the protein specified by the foreign DNA.

A subsequent generation of genetic engineering techniques that emerged in the early 21st century centred on gene editing. Gene editing, based on a technology known as CRISPR-Cas9, allows researchers to customize a living organism's genetic sequence by making very specific changes to its DNA. Gene editing has a wide array of applications, being used for the genetic modification of crop plants and livestock and of laboratory model organisms (e.g., mice). The correction of genetic errors associated with disease in animals suggests that gene editing has potential applications in gene therapy for humans.

Applications

Genetic engineering has advanced the understanding of many theoretical and practical aspects of gene function and organization. Through recombinant DNA techniques, bacteria have been created that are capable of synthesizing human insulin, human growth hormone, alpha interferon, a hepatitis B vaccine, and other medically useful substances. Plants may be genetically adjusted to enable them to fix nitrogen, and genetic diseases can possibly be corrected by replacing dysfunctional genes with normally functioning genes. Nevertheless, special concern has been focused on such achievements for fear that they might result in the introduction of unfavourable and possibly dangerous traits into microorganisms that were previously free of them—e.g., resistance to antibiotics, production of toxins, or a tendency to cause disease. Likewise, the application of gene editing in humans has raised ethical concerns, particularly regarding its potential use to alter traits such as intelligence and beauty.

Controversy

In 1980 the "new" microorganisms created by recombinant DNA research were deemed patentable, and in 1986 the U.S. Department of Agriculture approved the sale of the first living genetically altered organism—a virus, used as a pseudorabies vaccine, from which a single gene had been cut. Since then several hundred patents have been awarded for genetically altered bacteria and plants. Patents on genetically engineered and genetically modified organisms, particularly crops and other foods, however, were a contentious issue, and they remained so into the first part of the 21st century.

GENETICALLY MODIFIED ORGANISM

Genetically modified organism (GMO) is an organism whose genome has been engineered in the laboratory in order to favour the expression of desired physiological traits or the production of desired biological products. In conventional livestock production, crop farming, and even pet breeding, it has long been the practice to breed select individuals of a species in order to produce offspring that have desirable traits. In genetic modification, however, recombinant genetic technologies are employed to produce organisms whose genomes have been precisely altered at the molecular level, usually by the inclusion of genes from unrelated species of organisms that code for traits that would not be obtained easily through conventional selective breeding.

Genetically modified (GM) barley grown by researchers on a site belonging to Giessen University (Justus-Liebig-Universität) in Germany. The GM barley was investigated for its effects on soil quality.

GMOs are produced through using scientific methods that include recombinant DNA technology and reproductive cloning. In reproductive cloning, a nucleus is extracted from a cell of the individual to be cloned and is inserted into the enucleated cytoplasm of a host egg. The process results in the generation of an offspring that is genetically identical to the donor individual. The first animal produced by means of this cloning technique with a nucleus from an adult donor cell (as opposed to a donor embryo) was a sheep named Dolly, born in 1996. Since then a number of other animals, including pigs, horses, and dogs, have been generated by reproductive cloning technology. Recombinant DNA technology, on the other hand, involves the insertion of one or more individual genes from an organism of one species into the DNA (deoxyribonucleic acid) of another. Whole-genome replacement, involving the transplantation of one bacterial genome into the "cell body," or cytoplasm, of another microorganism, has been reported, although this technology is still limited to basic scientific applications.

GMOs produced through genetic technologies have become a part of everyday life, entering into society through agriculture, medicine, research, and environmental management. However, while GMOs have benefited human society in many ways, some disadvantages exist; therefore, the production of GMOs remains a highly controversial topic in many parts of the world.

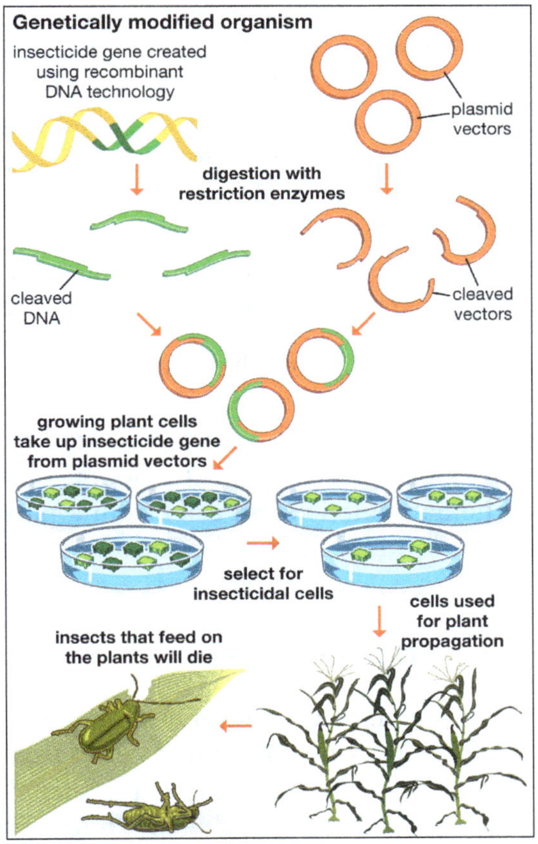

Genetically modified organisms are produced using scientific methods that include recombinant DNA technology.

GMOs in Medicine and Research

GMOs have emerged as one of the mainstays of biomedical research since the 1980s. For example, GM animal models of human genetic diseases enabled researchers to test novel therapies and to explore the roles of candidate risk factors and modifiers of disease outcome. GM microbes, plants, and animals also revolutionized the production of complex pharmaceuticals by enabling the generation of safer and cheaper vaccines and therapeutics. Pharmaceutical products range from recombinant hepatitis B vaccine produced by GM baker's yeast to injectable insulin (for diabetics) produced in GM Escherichia coli bacteria and to factor VIII (for hemophiliacs) and tissue plasminogen activator (tPA, for heart attack or stroke patients), both of which are produced in GM mammalian cells grown in laboratory culture. Furthermore, GM plants that produce "edible vaccines" are under development. An edible vaccine is an antigenic protein that is produced in the consumable parts of a plant (e.g., fruit) and absorbed into the bloodstream when the parts are eaten. Once absorbed into the body, the protein stimulates the immune system to produce antibodies against the pathogen from which the antigen was derived. Such vaccines could offer a safe, inexpensive, and painless way to provide vaccines, particularly in less-developed regions of the world, where the limited availability of refrigeration and sterile needles has been problematic for some traditional vaccines. Novel DNA vaccines may be useful in the struggle to prevent diseases that have proved resistant to traditional vaccination approaches, including HIV/AIDS, tuberculosis, and cancer.

Genetic modification of insects has become an important area of research, especially in the struggle to prevent parasitic diseases. For example, GM mosquitoes have been developed that express a small protein called SM1, which blocks entry of the malaria parasite, Plasmodium, into the mosquito's gut. This results in the disruption of the parasite's life cycle and renders the mosquito malaria-resistant. Introduction of these GM mosquitoes into the wild could help reduce transmission of the malaria parasite. In another example, male Aedes aegypti mosquitoes engineered with a method known as the sterile insect technique transmit a gene to their offspring that causes the offspring to die before becoming sexually mature. In field trials in a Brazil suburb, A. aegypti populations declined by 95 percent following the sustained release of sterile GM males.

Finally, genetic modification of humans via gene therapy is becoming a treatment option for diseases ranging from rare metabolic disorders to cancer. Coupling stem cell technology with recombinant DNA methods allows stem cells derived from a patient to be modified in the laboratory to introduce a desired gene. For example, a normal beta-globin gene may be introduced into the DNA of bone marrow-derived hematopoietic stem cells from a patient with sickle cell anemia; introduction of these GM cells into the patient could cure the disease without the need for a matched donor.

Role of GMOs in Environmental Management

Another application of GMOs is in the management of environmental issues. For example, some bacteria can produce biodegradable plastics, and the transfer of that ability to microbes that can be easily grown in the laboratory may enable the wide-scale "greening" of the plastics industry. In the early 1990s, Zeneca, a British company, developed a microbially produced biodegradable plastic called Biopol (polyhydroxyalkanoate, or PHA). The plastic was made with the use of a GM bacterium, Ralstonia eutropha, to convert glucose and a variety of organic acids into a flexible polymer. GMOs endowed with the bacterially encoded ability to metabolize oil and heavy metals may provide efficient bioremediation strategies.

Sociopolitical Relevance of GMOs

While GMOs offer many potential benefits to society, the potential risks associated with them have fueled controversy, especially in the food industry. Many skeptics warn about the dangers that GM crops may pose to human health. For example, genetic manipulation may potentially alter the allergenic properties of crops. Whether some GM crops, such as golden rice, deliver on the promise of improved health benefits is also unclear. The release of GM mosquitoes and other GMOs into the environment also raised concerns. More-established risks were associated with the potential spread of engineered crop genes to native flora and the possible evolution of insecticide-resistant "superbugs."

From the late 1990s, the European Union (EU) addressed such concerns by implementing strict GMO labeling laws. In the early 2000s, all GM foods and GM animal feeds in the EU were required to be labeled if they consisted of or contained GM products in a proportion greater than 0.9 percent. By contrast, in the United States, foods containing GM ingredients did not require special labeling, though the issue was hotly debated at national and state levels. Many opponents of GM products focused their arguments on unknown risks to food safety. However, despite the concerns of some consumer and health groups, especially in Europe, numerous scientific panels, including the U.S. Food and Drug Administration, concluded that consumption of GM foods was safe, even in cases involving GM foods with genetic material from very distantly related organisms.

The strict regulations on GM products in the EU have been a source of tension in agricultural trade. In the late 1990s, the EU declared a moratorium on the import and use of GM crops. However, the ban—which led to trade disputes with other countries, particularly the United States, where GM foods had been accepted openly—was considered unjustified by the World Trade Organization. In consequence, the EU implemented regulatory changes that allowed for the import of certain GM crops. Within Europe, however, only one GM crop, a type of insect-resistant corn (maize), was cultivated. Some countries, including certain African states, had likewise rejected GM products. Still other countries, such as Canada, China, Argentina, and Australia, had open policies on GM foods.

The use of GMOs in medicine and research has produced a debate that is more philosophical in nature. For example, while genetic researchers believe they are working to cure disease and ameliorate suffering, many people worry that current gene therapy approaches may one day be applied to produce "designer" children or to lengthen the natural human life span. Similar to many other technologies, gene therapy and the production and application of GMOs can be used to address and resolve complicated scientific, medical, and environmental issues, but they must be used wisely.

GENETICALLY MODIFIED ANIMAL

Genetically modified animals are animals that have been genetically modified for a variety of purposes including producing drugs, enhancing yields, increase resistance to disease, etc. The vast majority of genetically modified animals are at the research stage with the number close to entering the market remains small.

Production

The process of genetically engineering mammals is a slow, tedious, and expensive process. As with other genetically modified organisms (GMOs), first genetic engineers must isolated the gene they wish to insert into the host organism. This can be taken from a cell containing the gene or artificially synthesised. If the chosen gene or the donor organism's genome has been well studied it may already be accessible from a genetic library. The gene is then combined with other genetic elements, including a promoter and terminator region and usually a selectable marker.

A number of techniques are available for inserting the isolated gene into the host genome. With animals DNA is generally inserted into using microinjection, where it can be injected through the cell's nuclear envelope directly into the nucleus, or through the use of viral vectors. The first transgenic animals were produced by injecting viral DNA into embryos and then implanting the embryos in females. It is necessary to ensure that the inserted DNA is present in the embryonic stem cells. The embryo would develop and it would be hoped that some of the genetic material would be incorporated into the reproductive cells. Then researchers would have to wait until the animal reached breeding age and then offspring would be screened for presence of the gene in every cell, using PCR, Southern hybridization, and DNA sequencing.

New technologies are making genetic modifications easier and more precise. Gene targeting techniques, which creates double-stranded breaks and takes advantage on the cells natural homologous recombinationrepair systems, have been developed to target insertion to exact locations. Genome editing uses artificially engineered nucleases that create breaks at specific points. There are four families of engineered nucleases: meganucleases, zinc finger nucleases, transcription activator-like effector nucleases (TALENs), and the Cas9-guideRNA system (adapted from CRISPR). TALEN and CRISPR are the two most commonly used and each has its own advantages. TALENs have greater target specificity, while CRISPR is easier to design and more efficient. The development of the CRISPR-Cas9 gene editing system has effectively halved the amount of time needed to develop genetically modified animals.

Mammals

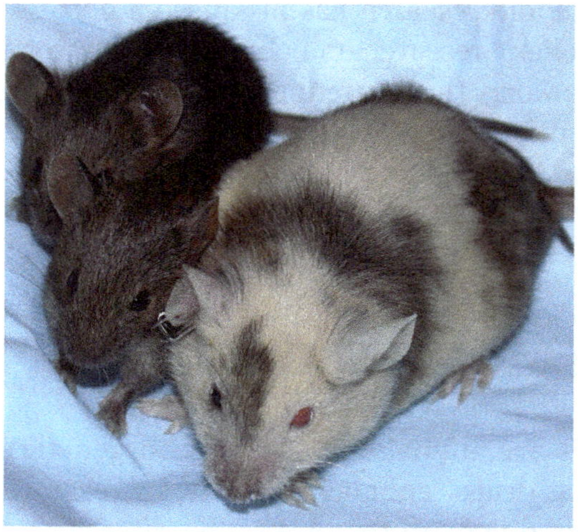

Some chimeras, like the blotched mouse shown, are created through genetic modification techniques like gene targeting.

GM mammals are created for research purposes, production of industrial or therapeutic products, agricultural uses or improving their health. There is also a market for creating genetically modified pets.

Medicine

Mammals are the best models for human disease, making genetic engineered ones vital to the discovery and development of cures and treatments for many serious diseases. Knocking out genes responsible for human genetic disorders allows researchers to study the mechanism of the disease and to test possible cures. Genetically modified micehave been the most common mammals used in biomedical research, as they are cheap and easy to manipulate. Pigs are also a good target as they have a similar body size and anatomical features, physiology, pathophysiological response and diet. Nonhuman primates are the most similar model organisms to humans, but there is less public acceptance towards using them as research animals. In 2009, scientists announced that they had successfully transferred a gene into a primate species (marmosets) and produced a stable line of breeding transgenic primates for the first time. Their first research target for these marmosets was Parkinson's disease, but they were also considering amyotrophic lateral sclerosis and Huntington's disease.

Human proteins expressed in mammals are more likely to be similar to their natural counterparts than those expressed in plants or microorganisms. Stable expression has been accomplished in sheep, pigs, rats and other animals. In 2009, the first human biological drug produced from such an animal, a goat., was approved. The drug, ATryn, is an anticoagulant which reduces the probability of blood clots during surgery or childbirth was extracted from the goat's milk. Human alpha-1-antitrypsin is another protein that is used in treating humans with this deficiency. Another area is in creating pigs with greater capacity for human organ transplants (xenotransplantation). Pigs have been genetically modified so that their organs can no longer carry retroviruses or have modifications to reduce the chance of rejection. Pig lungs from genetically modified pigs are being considered for transplantation into humans. There is even potential to create chimeric pigs that can carry human organs.

Livestock

Livestock are modified with the intention of improving economically important traits such as growth-rate, quality of meat, milk composition, disease resistance and survival. Animals have been engineered to grow faster, be healthier and resist diseases. Modifications have also improved the wool production of sheep and udder health of cows.

Goats have been genetically engineered to produce milk with strong spiderweb-like silk proteins in their milk.The goat gene sequence has been modified, using fresh umbilical cords taken from kids, in order to code for the human enzyme lysozyme. Researchers wanted to alter the milk produced by the goats, to contain lysozyme in order to fight off bacteria causing diarrhea in humans.

Enviropig was a genetically enhanced line of Yorkshire pigs in Canada created with the capability of digesting plant phosphorus more efficiently than conventional Yorkshire pigs. The A transgene construct consisting of a promoter expressed in the murine parotid gland and the *Escherichia coli* phytase gene was introduced into the pig embryo by pronuclear microinjection. This caused the pigs to produce the enzyme phytase, which breaks down the indigestible phosphorus, in their saliva. As a result they excrete 30 to 70% less phosphorus in manure depending upon the age and diet. The lower concentrations of phosphorus in surface runoff reduces algal growth, because phosphorus is the limiting nutrient for algae. Because algae consume large amounts of oxygen, excessive growth can result in dead zones for fish. Funding for the Enviropig program ended in April 2012, and as no new partners were found the pigs were killed. However, the genetic material will be stored at the Canadian Agricultural Genetics Repository Program. In 2006, a pig was engineered to produce omega-3 fatty acids through the expression of a roundworm gene.

In 1990, the world's first transgenic bovine, Herman the Bull, was developed. Herman was genetically engineered by micro-injected embyonic cells with the human gene coding for lactoferrin. The Dutch Parliament changed the law in 1992 to allow Herman to reproduce. Eight calves were born in 1994 and all calves inherited the lactoferrin gene. With subsequent sirings, Herman fathered a total of 83 calves. Dutch law required Herman to be slaughtered at the conclusion of the experiment. However the Dutch Agriculture Minister at the time, Jozias van Aartsen, granted him a reprieve provided he did not have more offspring after public and scientists rallied to his defence.Together with cloned cows named Holly and Belle, he lived out his retirement at Naturalis, the National Museum of Natural History in Leiden. In 2004, Herman was euthanised by veterinarians from the University of Utrecht because he suffered from osteoarthritis. At the time of his death Herman was one of the oldest bulls in the Netherlands. Herman's hide has been preserved and mounted by taxidermists and is permanently on display in Naturalis. They say that he represents the start of a new era in the way man deals with nature, an icon of scientific progress, and the subsequent public discussion of these issues.

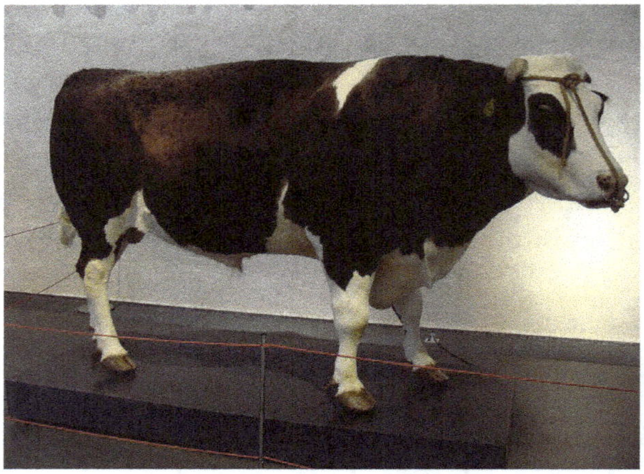

Herman the Bull on display in Naturalis Biodiversity Center.

Researchers have developed GM dairy cattle to grow without horns (sometimes referred to as "polled") which can cause injuries to farmers and other animals. DNA was taken from the genome of Red Angus cattle, which is known to suppress horn growth, and inserted into cells taken from an elite Holstein bull called "Randy". Each of the progeny will be a clone of Randy, but without his horns, and their offspring should also be hornless. In 2011, Chinese scientists generated dairy cows genetically engineered with genes from human beings to produce milk that would be the same as human breast milk. This could potentially benefit mothers who cannot produce breast milk but want their children to have breast milk rather than formula. The researchers claim these transgenic cows to be identical to regular cows. Two months later, scientists from Argentina presented Rosita, a transgenic cow incorporating two human genes, to produce milk with similar properties as human breast milk. In 2012, researchers from New Zealand also developed a genetically engineered cow that produced allergy-free milk.

Conservation

Genetically modification with a myxoma virus has been proposed to conserve European wild rabbits in the Iberian peninsula and to help regulate them in Australia. To protect the Iberian species from viral diseases, the myxoma virus was genetically modified to immunize the rabbits, while in Australia the same myxoma virus was genetically modified to lower fertility in the Australian rabbit population. There have also been suggestions that genetic engineering could be used to bring animals back from extinction. It involves changing the genome of a close living relative to resemble the extinct one and is currently being attempted with the passenger pigeon.Genes associated with the woolly mammoth have been added to the genome of an African Elephant, although the lead researcher says he has no intention of using live elephants.

Humans

Gene therapy, uses genetically modified viruses to deliver genes which can cure disease in humans. Although gene therapy is still relatively new, it has had some successes. It has been used to treat genetic disorders such as severe combined immunodeficiency, and Leber's congenital amaurosis. Treatments are also being developed for a range of other currently incurable diseases, such as cystic fibrosis, sickle cell anemia, Parkinson's disease, cancer, diabetes, heart disease and muscular dystrophy. These treatments only effect somatic cells, meaning any changes would not be inheritable. Germline gene therapy results in any change being inheritable, which has raised concerns within the scientific community. In 2015, CRISPR was used to edit the DNA of non-viable human embryos. In November 2018, He Jiankui announced that he had edited the genomes of two human embryos, to attempt to disable the *CCR5* gene, which codes for a receptor that HIV uses to enter cells. He said that twin girls, Lulu and Nana, had been born a few weeks earlier and that they carried functional copies of CCR5 along with disabled CCR5 (mosaicism) and were still vulnerable to HIV. The work was widely condemned as unethical, dangerous, and premature.

Fish

Genetically modified fish are used for scientific research, as pets and as a food source. Aquaculture is a growing industry, currently providing over half the consumed fish worldwide. Through genetic engineering It is possible to increase growth rates, reduce food intake, remove allergenic properties, increase cold tolerance and provide disease resistance.

Detecting Pollution

Fish can also be used to detect aquatic pollution or function as bioreactors. Several groups have been developing zebrafish to detect pollution by attaching fluorescent proteins to genes activated by the presence of pollutants. The fish will then glow and can be used as environmental sensors.

Pets

The GloFish is a brand of genetically modified fluorescent zebrafish with bright red, green, and orange fluorescent color. It was originally developed by one of the groups to detect pollution, but is now part of the ornamental fish trade, becoming the first genetically modified animal to become publicly available as a pet when it was introduced for sale in 2003.

Growth

GM fish have been developed with promoters driving an over-production of "all fish" growth hormone for use in the aquaculture industry to increase the speed of development and potentially reduce fishing pressure on wild stocks. This has resulted in dramatic growth enhancement in several species, including salmon, troutand tilapia.

AquaBounty Technologies have produced a salmon that can mature in half the time as wild salmon. The fish is an Atlantic salmon with a Chinook salmon (*Oncorhynchus tshawytscha*) gene inserted. This allows the fish to produce growth hormones all year round compared to the wild-type fish that produces the hormone for only part of the year. The fish also has a second gene inserted from the eel-like ocean pout that acts like an "on" switch for the hormone. Pout also have antifreeze proteins in their blood, which allow the GM salmon to survive near-freezing waters and continue their development. The wild-type salmon takes 24 to 30 months to reach market size (4–6 kg) whereas the producers of the GM salmon say it requires only 18 months for the GM fish to achieve this. In November 2015, the FDA of the USA approved the AquAdvantage salmon for commercial production, sale and consumption, the first non-plant GMO food to be commerialised.

AquaBounty say that to prevent the genetically modified fish inadvertently breeding with wild salmon, all the fish will be female and reproductively sterile, although a small percentage of the females may remain fertile.Some opponents of the GM salmon have dubbed it the "Frankenfish".

Population Control

Due to their significance to human health, scientists are looking at ways to control mosquitoes through genetic engineering. Malaria-resistant mosquitoes have been developed in the laboratory. by inserting a gene that reduces the development of the malaria parasite and then use homing endonucleases to rapidly spread that gene throughout the male population (known as a gene drive). This has been taken further by swapping it for a lethal gene. In trials the populations of *Aedes aegypti* mosquitoes, the single most important carrier of dengue fever and Zika virus, were reduced by between 80% and by 90%. Another approach is to use the sterile insect technique, whereby males genetically engineered to be sterile out compete viable males, to reduce population numbers.

Other insect pests that make attractive targets are moths. Diamondback moths cause US$4 to $5 billion of damage a year worldwide. The approach is similar to the mosquitoes, where males transformed with a gene that prevents females from reaching maturity will be released. They underwent field trials in 2017. Genetically modified moths have previously been released in field trials. A strain of pink bollworm that were sterilised with radiation were genetically engineered to express a red fluorescent protein making it easier for researchers to monitor them.

Industry

Silkworm, the larvae stage of *Bombyx mori,* is an economically important insect in sericulture. Scientists are developing strategies to enhance silk quality and quantity. There is also potential to use the silk producing machinery to make other valuable proteins. Proteins expressed by silkworms include; human serum albumin, human collagen α-chain, mouse monoclonal antibody and N-glycanase. Silkworms have been created that produce spider silk, a stronger but extremely difficult to harvest silk, and even novel silks.

Birds

Attempts to produce genetically modified birds began before 1980. Chickens have been genetically modified for a variety of purposes. This includes studying embryo development, preventing the transmission of bird flu and providing evolutionary insights using reverse engineering to recreate dinosaur-like phenotypes. A GM chicken that produces the drug Kanuma, an enzyme that treats a rare condition, in its egg passed regulatory approval in 2015.

Disease Control

One potential use of GM birds could be to reduce the spread of avian disease. Researchers at Roslin Institutehave produced a strain of GM chickens (*Gallus gallus domesticus*) that does not transmit avian flu to other birds; however, these birds are still susceptible to contracting it. The genetic modification is an RNA molecule that prevents the virus reproduction by mimicking the region of the flu virus genome that controls replication. It is referred to as a "decoy" because it diverts the flu virus enzyme, the polymerase, from functions that are required for virus replication.

Evolutionary Insights

A team of geneticists led by University of Montana paleontologist Jack Horner is seeking to modify a chicken to express several features present in ancestral maniraptorans but absent in modern birds, such as teeth and a long tail, creating what has been dubbed a 'chickenosaurus'. Parallel projects have produced chicken embryos expressing dinosaur-like skull, leg, and foot anatomy.

Amphibians

Genetically modified frogs, in particular *Xenopus laevis* and *Xenopus tropicalis*, are used in development biology. GM frogs can also being used as pollution sensors, especially for endocrine disrupting chemicals.There are proposals to use genetic engineering to control cane toads in Australia.

Nematodes

The nemotode *Caenorhabditis elegans* is one of the major model organisms for researching molecular biology. RNA interference (RNAi) was discovered in *C elegans* and could be induced by simply feeding them bacteria modified to express double stranded RNA. It is also relatively easy to produce stable transgenic nemotodes and this along with RNAi are the major tools used in studying their genes. The most common use of transgenic nematodes has been studying gene expression and localisation by attaching reporter genes. Transgenes can also be combined with RNAi to rescue phenotypes, altered to study gene function, imaged in real time as the cells develop or used to control expression for different tissues or developmental stages. Transgenic nematodes have been used to study viruses, toxicology, and diseasesand to detect environmental pollutants.

Other

Systems have been developed to create transgenic organisms in a wide variety of other animals. The gene responsible for Albinism in sea cucumbers has been found and used to engineer white sea cucumbers, a rare delicacy. The technology also opens the way to investigate the genes responsible for some of the cucumbers more unusual traits, including hibernating in summer, eviscerating their intestines, and dissolving their bodies upon death. Flatworms have the ability to regenerate themselves from a single cell. Until 2017 there was no effective way to transform them, which hampered research. By using microinjection and radiation scientist have now created the first genetically modified flatworms. The bristle worm, a marine annelid, has been modified. It is of interest due to its reproductive cycle being synchronised with lunar phases, regeneration capacity and slow evolution rate. Cnidaria such as *Hydra* and the sea anemone *Nematostella vectensis* are attractive model organisms to study the evolution of immunity and certain developmental processes. Other organisms that have been genetically modified include snails, geckos, turtles, crayfish, oysters, shrimp, clams, abalone and sponges.

GENETICALLY MODIFIED FOOD

Genetically modified foods (GM foods), also known as genetically engineered foods (GE foods), or bioengineered foods are foods produced from organisms that have had changes introduced into their DNA using the methods of genetic engineering. Genetic engineering techniques allow for the introduction of new traits as well as greater control over traits when compared to previous methods, such as selective breedingand mutation breeding.

Commercial sale of genetically modified foods began in 1994, when Calgene first marketed its unsuccessful Flavr Savr delayed-ripening tomato. Most food modifications have primarily focused on cash crops in high demand by farmers such as soybean, corn, canola, and cotton. Genetically modified cropshave been engineered for resistance to pathogens and herbicidesand for better nutrient profiles. GM livestock have been developed, although, as of November 2013, none were on the market.

There is a scientific consensus that currently available food derived from GM crops poses no greater risk to human health than conventional food, but that each GM food needs to be tested on a case-by-case basis before introduction. Nonetheless, members of the public are much less likely than scientists to perceive GM foods as safe. The legal and regulatory status of GM foods varies by country, with some nations banning or restricting them, and others permitting them with widely differing degrees of regulation.

However, there are ongoing public concerns related to food safety, regulation, labelling, environmental impact, research methods, and the fact that some GM seeds, along with all new plant varieties, are subject to plant breeders' rights owned by corporations.

Genetically modified foods are foods produced from organisms that have had changes introduced into their DNA using the methods of genetic engineering as opposed to traditional cross breeding. In the U.S., the Department of Agriculture (USDA) and the Food and Drug Administration (FDA) favor the use of the term *genetic engineering* over *genetic modification* as being more precise; the USDA defines *genetic modification* to include "genetic engineering or other more traditional methods".

According to the World Health Organization, "genetically modified organisms (GMOs) can be defined as organisms (i.e. plants, animals or microorganisms) in which the genetic material (DNA) has been altered in a way that does not occur naturally by mating and/or natural recombination. The technology is often called 'modern biotechnology' or 'gene technology', sometimes also 'recombinant DNA technology' or 'genetic engineering'. Foods produced from or using GM organisms are often referred to as GM foods."

Process

Genetically engineered organisms are generated and tested in the laboratory for desired qualities. The most common modification is to add one or more genes to an organism's genome. Less commonly, genes are removed or their expression is increased or silenced or the number of copies of a gene is increased or decreased.

Once satisfactory strains are produced, the producer applies for regulatory approval to field-test them, called a "field release". Field-testing involves cultivating the plants on farm fields or growing animals in a controlled environment. If these field tests are successful, the producer applies for regulatory approval to grow and market the crop. Once approved, specimens (seeds, cuttings, breeding pairs, etc.) are cultivated and sold to farmers. The farmers cultivate and market the new strain. In some cases, the approval covers marketing but not cultivation.

According to the USDA, the number of field releases for genetically engineered organisms has grown from four in 1985 to an average of about 800 per year. Cumulatively, more than 17,000 releases had been approved through September 2013.

Crops

Fruits and Vegetables

Three views of a papaya, cultivar "Sunset", which was genetically modified to create the cultivar 'SunUp', which is resistant to Papaya ringspot virus.

Papaya was genetically modified to resist the ringspot virus(PSRV). "SunUp" is a transgenic red-fleshed Sunset papaya cultivar that is homozygous for the coat protein gene PRSV; "Rainbow" is a yellow-fleshed F1 hybrid developed by crossing 'SunUp' and nontransgenic yellow-fleshed "Kapoho". *The New York Times* stated, "in the early 1990s, Hawaii's papaya industry was facing disaster because of the deadly papaya ringspot virus. Its single-handed savior was a breed engineered to be resistant to the virus. Without it, the state's papaya industry would have collapsed. Today, 80% of Hawaiian papaya is genetically engineered, and there is still no conventional or organic method to control ringspot virus." The GM cultivar was approved in 1998. In China, a transgenic PRSV-resistant papaya was developed by South China Agricultural University and was first approved for commercial planting in 2006; as of 2012 95% of the papaya grown in Guangdong province and 40% of the papaya grown in Hainan province was genetically modified. In Hong Kong, where there is an exemption on growing and releasing any varieties of GM papaya, more than 80% of grown and imported papayas were transgenic.

The New Leaf potato, a GM food developed using naturally occurring bacteria found in the soil known as *Bacillus thuringiensis* (Bt), was made to provide in-plant protection from the yield-robbing Colorado potato beetle. The New Leaf potato, brought to market by Monsanto in the late 1990s, was developed for the fast food market. It was withdrawn in 2001 after retailers rejected it and food processors ran into export problems.

Plums genetically engineered for resistance to plum pox, a disease carried by aphids.

As of 2005, about 13% of the Zucchini (a form of squash) grown in the US was genetically modified to resist three viruses; that strain is also grown in Canada.

In 2011, BASF requested the European Food Safety Authority's approval for cultivation and marketing of its Fortuna potato as feed and food. The potato was made resistant to late blight by adding resistant genes blb1 and blb2 that originate from the Mexican wild potato Solanum bulbocastanum. In 2013, BASF withdrew its application.

In 2013, the USDA approved the import of a GM pineapple that is pink in color and that "overexpresses" a gene derived from tangerines and suppress other genes, increasing production of lycopene. The plant's flowering cycle was changed to provide for more uniform growth and quality. The fruit "does not have the ability to propagate and persist in the environment once they have been harvested", according to USDA APHIS. According to Del Monte's submission, the pineapples are commercially grown in a "monoculture" that prevents seed production, as the plant's flowers aren't exposed to compatible pollen sources. Importation into Hawaii is banned for "plant sanitation" reasons.

In 2014, the USDA approved a genetically modified potato developed by J. R. Simplot Company that contained ten genetic modifications that prevent bruising and produce less acrylamide when fried. The modifications eliminate specific proteins from the potatoes, via RNA interference, rather than introducing novel proteins.

In 2015 Arctic Apples were approved by the USDA, becoming the first genetically modified apple approved for sale in the US. Gene silencing is used to reduce the expression of polyphenol oxidase (PPO), thus preventing the fruit from browning.

Corn

Corn used for food and ethanol has been genetically modified to tolerate various herbicides and to express a protein from *Bacillus thuringiensis* (Bt) that kills certain insects. About 90% of the corn grown in the US was genetically modified in 2010. In the US in 2015, 81% of corn acreage contained the Bt trait and 89% of corn acreage contained the glyphosate-tolerant trait. Corn can be processed into grits, meal and flour as an ingredient in pancakes, muffins, doughnuts, breadings and batters, as well as baby foods, meat products, cereals and some fermented products. Corn-based masa flour and masa dough are used in the production of taco shells, corn chips and tortillas.

Soy

Genetically modified soybean has been modified to tolerate herbicides and produce healthier oils. In 2015, 94% of soybean acreage in the U.S. was genetically modified to be glyphosate-tolerant.

Wheat

As of December 2017, genetically modified wheat has been evaluated in field trials, but has not been released commercially.

Derivative Products

Corn Starch and Starch Sugars and Including Syrups

Starch or amylum is a polysaccharide produced by all green plants as an energy store. Pure starch is a white, tasteless and odourless powder. It consists of two types of molecules: the linear and helical amylose and the branched amylopectin. Depending on the plant, starch generally contains 20 to 25% amylose and 75 to 80% amylopectin by weight.

Starch can be further modified to create modified starch for specific purposes, including creation of many of the sugars in processed foods. They include:

- Maltodextrin, a lightly hydrolyzed starch product used as a bland-tasting filler and thickener.
- Various glucose syrups, also called corn syrups in the US, viscous solutions used as sweeteners and thickeners in many kinds of processed foods.
- Dextrose, commercial glucose, prepared by the complete hydrolysis of starch.
- High fructose syrup, made by treating dextrose solutions with the enzyme glucose isomerase, until a substantial fraction of the glucose has been converted to fructose.
- Sugar alcohols, such as maltitol, erythritol, sorbitol, mannitol and hydrogenated starch hydrolysate, are sweeteners made by reducing sugars.

Lecithin

Lecithin is a naturally occurring lipid. It can be found in egg yolks and oil-producing plants. It is an emulsifier and thus is used in many foods. Corn, soy and safflower oil are sources of lecithin, though the majority of lecithin commercially available is derived from soy. Sufficiently processed lecithin is often undetectable with standard testing practices. According to the FDA, no evidence shows or suggests hazard to the public when lecithin is used at common levels. Lecithin added to foods amounts to only 2 to 10 percent of the 1 to 5 g of phosphoglycerides consumed daily on average. Nonetheless, consumer concerns about GM food extend to such products. This concern led to policy and regulatory changes in Europe in 2000, when Regulation (EC) 50/2000 was passed which required labelling of food containing additives derived from GMOs, including lecithin. Because of the difficulty of detecting the origin of derivatives like lecithin with current testing practices, European regulations require those who wish to sell lecithin in Europe to employ a comprehensive system of Identity preservation(IP).

Sugar

The US imports 10% of its sugar, while the remaining 90% is extracted from sugar beet and sugarcane. After deregulation in 2005, glyphosate-resistant sugar beet was extensively adopted in the United States. 95% of beet acres in the US were planted with glyphosate-resistant seed in 2011. GM sugar beets are approved for cultivation in the US, Canada and Japan; the vast majority are grown in the US. GM beets are approved for import and consumption in Australia, Canada, Colombia, EU, Japan, Korea, Mexico, New Zealand, Philippines, the Russian Federation and Singapore. Pulp from the refining process is used as animal feed. The sugar produced from GM sugar beets contains no DNA or protein – it is just sucrose that is chemically indistinguishable from sugar produced from non-GM sugar beets. Independent analyses conducted by internationally recognized laboratories found that sugar from Roundup Ready sugar beets is identical to the sugar from comparably grown conventional (non-Roundup Ready) sugar beets.

Vegetable Oil

Most vegetable oil used in the US is produced from GM crops canola, corn, cotton and soybeans. Vegetable oil is sold directly to consumers as cooking oil, shortening and margarine and is used in prepared foods. There is a vanishingly small amount of protein or DNA from the original crop in vegetable oil.Vegetable oil is made of triglycerides extracted from plants or seeds and then refined and may be further processed via hydrogenation to turn liquid oils into solids. The refining process removes all, or nearly all non-triglyceride ingredients. Medium-chain triglycerides (MCTs) offer an alternative to conventional fats and oils. The length of a fatty acid influences its fat absorption during the digestive process. Fatty acids in the middle position on the glycerol molecules appear to be absorbed more easily and influence metabolism more than fatty acids on the end positions. Unlike ordinary fats, MCTs are metabolized like carbohydrates. They have exceptional oxidative stability, and prevent foods from turning rancid readily.

Other Uses

Animal Feed

Livestock and poultry are raised on animal feed, much of which is composed of the leftovers from processing crops, including GM crops. For example, approximately 43% of a canola seed is oil. What remains after oil extraction is a meal that becomes an ingredient in animal feed and contains canola protein. Likewise, the bulk of the soybean crop is grown for oil and meal. The high-protein defatted and toasted soy meal becomes livestock feed and dog food. 98% of the US soybean crop goes for livestock feed. In 2011, 49% of the US maize harvest was used for livestock feed (including the percentage of waste from distillers grains)."Despite methods that are becoming more and more sensitive, tests have not yet been able to establish a difference in the meat, milk, or eggs of animals depending on the type of feed they are fed. It is impossible to tell if an animal was fed GM soy just by looking at the resulting meat, dairy, or egg products. The only way to verify the presence of GMOs in animal feed is to analyze the origin of the feed itself."

A 2012 literature review of studies evaluating the effect of GM feed on the health of animals did not find evidence that animals were adversely affected, although small biological differences were occasionally found. The studies included in the review ranged from 90 days to two years, with several of the longer studies considering reproductive and intergenerational effects.

Enzymes produced by genetically modified microorganisms are also integrated into animal feed to enhance availability of nutrients and overall digestion. These enzymes may also provide benefit to the gut microbiome of an animal, as well as hydrolyse antinutritional factors present in the feed.

Proteins

Rennet is a mixture of enzymes used to coagulate milk into cheese. Originally it was available only from the fourth stomach of calves, and was scarce and expensive, or was available from microbial sources, which often produced unpleasant tastes. Genetic engineering made it possible to extract rennet-producing genes from animal stomachs and insert them into bacteria, fungi or yeasts to make them produce chymosin, the key enzyme. The modified microorganism is killed after fermentation. Chymosin is isolated from the fermentation broth, so that the Fermentation-Produced Chymosin (FPC) used by cheese producers has an amino acid sequence that is identical to bovine rennet. The majority of the applied chymosin is retained in the whey. Trace quantities of chymosin may remain in cheese.

FPC was the first artificially produced enzyme to be approved by the US Food and Drug Administration. FPC products have been on the market since 1990 and as of 2015 had yet to be surpassed in commercial markets. In 1999, about 60% of US hard cheese was made with FPC. Its global market share approached 80%. By 2008, approximately 80% to 90% of commercially made cheeses in the US and Britain were made using FPC.

In some countries, recombinant (GM) bovine somatotropin (also called rBST, or bovine growth hormone or BGH) is approved for administration to increase milk production. rBST may be present in milk from rBST treated cows, but it is destroyed in the digestive system and even if directly injected into the human bloodstream, has no observable effect on humans. The FDA, World Health Organization, American Medical Association, American Dietetic Association and the National Institutes of Health have independently stated that dairy products and meat from rBST-treated cows are safe for human consumption. However, on 30 September 2010, the United States Court of Appeals, Sixth Circuit, analyzing submitted evidence, found a "compositional difference" between milk from rBGH-treated cows and milk from untreated cows. The court stated that milk from rBGH-treated cows has: increased levels of the hormone Insulin-like growth factor 1 (IGF-1); higher fat content and lower protein content when produced at certain points in the cow's lactation cycle; and more somatic cell counts, which may "make the milk turn sour more quickly".

Livestock

Genetically modified livestock are organisms from the group of cattle, sheep, pigs, goats, birds, horses and fish kept for human consumption, whose genetic material (DNA) has been altered using genetic engineeringtechniques. In some cases, the aim is to introduce a new trait to the animals which does not occur naturally in the species, i.e. transgenesis.

A 2003 review published on behalf of Food Standards Australia New Zealand examined transgenic experimentation on terrestrial livestock species as well as aquatic species such as fish and shellfish. The review examined the molecular techniques used for experimentation as well as techniques for tracing the transgenes in animals and products as well as issues regarding transgene stability.

Some mammals typically used for food production have been modified to produce non-food products, a practice sometimes called Pharming.

Salmon

A GM salmon, awaiting regulatory approval since 1997, was approved for human consumption by the American FDA in November 2015, to be raised in specific land-based hatcheries in Canada and Panama.

Health and Safety

There is a scientific consensus that currently available food derived from GM crops poses no greater risk to human health than conventional food, but that each GM food needs to be tested on a case-by-case basis before introduction. Nonetheless, members of the public are much less likely than scientists to perceive GM foods as safe. The legal and regulatory status of GM foods varies by country, with some nations banning or restricting them, and others permitting them with widely differing degrees of regulation.

Opponents claim that long-term health risks have not been adequately assessed and propose various combinations of additional testing, labeling or removal from the market. The advocacy group European Network of Scientists for Social and Environmental Responsibility (ENSSER), disputes the claim that "science" supports the safety of current GM foods, proposing that each GM food must be judged on case-by-case basis.

Testing

The legal and regulatory status of GM foods varies by country, with some nations banning or restricting them, and others permitting them with widely differing degrees of regulation. Countries such as the United States, Canada, Lebanon and Egypt use *substantial equivalence* to determine if further testing is required, while many countries such as those in the European Union, Brazil and China only authorize GMO cultivation on a case-by-case basis. In the U.S. the FDA determined that GMO's are "Generally Recognized as Safe" (GRAS) and therefore do not require additional testing if the GMO product is substantially equivalent to the non-modified product. If new substances are found, further testing may be required to satisfy concerns over potential toxicity, allergenicity, possible gene transfer to humans or genetic outcrossing to other organisms.

Regulation

Government regulation of GMO development and release varies widely between countries. Marked differences separate GMO regulation in the U.S. and GMO regulation in the European Union. Regulation also varies depending on the intended product's use. For example, a crop not intended for food use is generally not reviewed by authorities responsible for food safety.

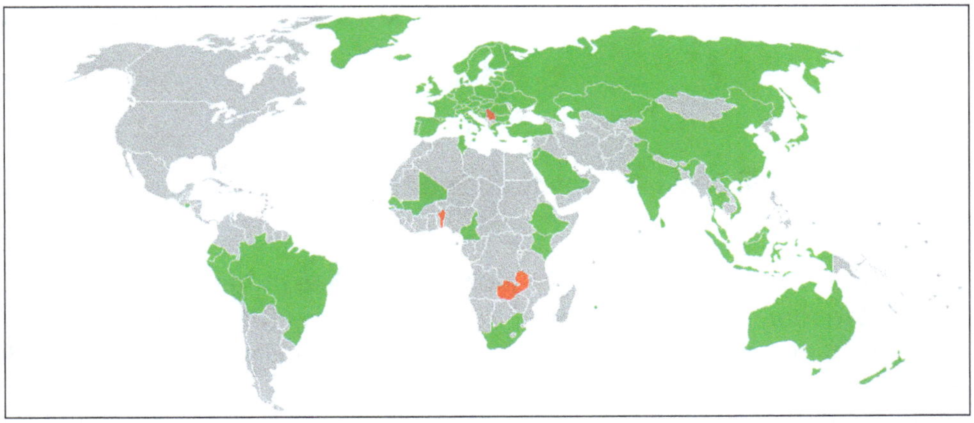

Green: Mandatory labeling required; Red: Ban on import and cultivation of genetically engineered food.

United States Regulations

In the U.S., three government organizations regulate GMOs. The FDAchecks the chemical composition of organisms for potential allergens. The United States Department of Agriculture (USDA) supervises field testing and monitors the distribution of GM seeds. The United States Environmental Protection Agency (EPA) is responsible for monitoring pesticide usage, including plants modified to contain proteins toxic to insects. Like USDA, EPA also oversees field testing and the distribution of crops that have had contact with pesticides to ensure environmental safety. In 2015 the Obama administration announced that it would update the way the government regulated GM crops.

In 1992 FDA published "Statement of Policy: Foods derived from New Plant Varieties". This statement is a clarification of FDA's interpretation of the Food, Drug, and Cosmetic Act with respect to foods produced from new plant varieties developed using recombinant deoxyribonucleic acid (rDNA) technology. FDA encouraged developers to consult with the FDA regarding any bioengineered foods in development. The FDA says developers routinely do reach out for consultations. In 1996 FDA updated consultation procedures.

The StarLink corn recalls occurred in the autumn of 2000, when over 300 food products were found to contain a genetically modified corn that had not been approved for human consumption. It was the first-ever recall of a genetically modified food.

Labeling

As of 2015, 64 countries require labeling of GMO products in the marketplace. US and Canadian national policy is to require a label only given significant composition differences or documented health impacts, although some individual US states (Vermont, Connecticut and Maine) enacted laws requiring them. In July 2016, Public Law 114-214 was enacted to regulate labeling of GMO food on a national basis.

In some jurisdictions, the labeling requirement depends on the relative quantity of GMO in the product. A study that investigated voluntary labeling in South Africa found that 31% of products labeled as GMO-free had a GM content above 1.0%.

In the European Union all food (including processed food) or feed that contains greater than 0.9% GMOs must be labelled.

Detection

Testing on GMOs in food and feed is routinely done using molecular techniques such as PCR and bioinformatics.

In 2010, the extraction and detection of DNA along a complete industrial soybean oil processing chain was described to monitor the presence of Roundup Ready (RR) soybean: "The amplification of soybean lectin gene by end-point polymerase chain reaction (PCR) was successfully achieved in all the steps of extraction and refining processes, until the fully refined soybean oil. The amplification of RR soybean by PCR assays using event-specific primers was also achieved for all the extraction and refining steps, except for the intermediate steps of refining (neutralisation, washing and bleaching) possibly due to sample instability. The real-time PCR assays using specific probes confirmed all the results and proved that it is possible to detect and quantify genetically modified organisms in the fully refined soybean oil. To our knowledge, this has never been reported before and represents an important accomplishment regarding the traceability of genetically modified organisms in refined oils."

According to Thomas Redick, detection and prevention of cross-pollination is possible through the suggestions offered by the Farm Service Agency (FSA) and Natural Resources Conservation Service (NRCS). Suggestions include educating farmers on the importance of coexistence, providing farmers with tools and incentives to promote coexistence, conduct research to understand and monitor gene flow, provide assurance of quality and diversity in crops, provide compensation for actual economic losses for farmers.

Controversy

The genetically modified foods controversy consists of a set of disputes over the use of food made from genetically modified crops. The disputes involve consumers, farmers, biotechnology companies, governmental regulators, non-governmental organizations, environmental and political activists and scientists. The major disagreements include whether GM foods can be safely consumed, harm the environment and/or are adequately tested and regulated. The objectivity of scientific research and publications has been challenged. Farming-related disputes include the use and impact of pesticides, seed production and use, side effects on non-GMO crops/farms, and potential control of the GM food supply by seed companies.

The conflicts have continued since GM foods were invented. They have occupied the media, the courts,local, regional, national governments, and international organizations.

GENETICALLY MODIFIED CROPS

Genetically modified crops (GM crops) are plants used in agriculture, the DNA of which has been modified using genetic engineering methods. In most cases, the aim is to introduce a new trait to the plant which does not occur naturally in the species. Examples in food crops include resistance to certain pests, diseases, environmental conditions, reduction of spoilage, resistance to chemical treatments (e.g. resistance to a herbicide), or improving the nutrient profile of the crop. Examples in non-food crops include production of pharmaceutical agents, biofuels, and other industrially useful goods, as well as for bioremediation.

Farmers have widely adopted GM technology. Acreage increased from 1.7 million hectares in 1996 to 185.1 million hectares in 2016, some 12% of global cropland. As of 2016, major crop (soybean, maize, canola and cotton) traits consist of herbicide tolerance (95.9 million hectares) insect resistance (25.2 million hectares), or both (58.5 million hectares). In 2015, 53.6 million ha of GM maize were under cultivation (almost 1/3 of the maize crop). GM maize outperformed its predecessors: yield was 5.6 to 24.5% higher with less mycotoxins (−28.8%), fumonisin (−30.6%) and thricotecens (−36.5%). Non-target organisms were unaffected, except for *Braconidae*, represented by a parasitoid of European corn borer, the target of *Lepidoptera* active Bt maize. Biogeochemical parameters such as lignin content did not vary, while biomass decomposition was higher.

A 2014 meta-analysis concluded that GM technology adoption had reduced chemical pesticide use by 37%, increased crop yields by 22%, and increased farmer profits by 68%. This reduction in pesticide use has been ecologically beneficial, but benefits may be reduced by overuse. Yield gains and pesticide reductions are larger for insect-resistant crops than for herbicide-tolerant crops. Yield and profit gains are higher in developing countries than in developed countries.

There is a scientific consensus that currently available food derived from GM crops poses no greater risk to human health than conventional food, but that each GM food needs to be tested on a case-by-case basis before introduction. Nonetheless, members of the public are much less likely than scientists to perceive GM foods as safe. The legal and regulatory status of GM foods varies by country, with some nations banning or restricting them, and others permitting them with widely differing degrees of regulation.

However, opponents have objected to GM crops on grounds including environmental impacts, food safety, whether GM crops are needed to address food needs, whether they are sufficiently accessible to farmers in developing countries and concerns over subjecting crops to intellectual property law. Safety concerns led 38 countries, including 19 in Europe, to officially prohibit their cultivation.

Nature and Traditional Agriculture

Multiple natural mechanisms allow gene flow from one species to another. These occur in nature on a large scale – for example, it is one mechanism for the development of antibiotic resistance in bacteria. This is facilitated by transposons, retrotransposons, retroviruses, proviruses and other mobile genetic elements that naturally translocate DNA to new loci in a genome. Movement occurs over an evolutionary time scale.

Traditional crop breeders have long introduced foreign germplasm into crops by creating novel crosses. A hybridcereal grain was created in 1875, by crossing wheat and rye. Since then traits including dwarfing genes and rust resistance have been introduced in that manner. Plant tissue culture and deliberate mutations have enabled humans to alter the makeup of plant genomes.

Direct Manipulation

The term genetic engineering is applied to genetic modifications made using biotechnology. The first plant produced in that way came in 1982, an antibiotic-resistant tobacco plant. The first field trials occurred in France and the USA in 1986, using tobacco plants engineered for herbicide resistance. In 1987, Plant Genetic Systems (Ghent, Belgium), founded by Marc Van Montagu and Jeff Schell, was the first company to genetically engineer insect-resistant (tobacco) plants by incorporating genes that produced insecticidal proteins from *Bacillus thuringiensis* (Bt).

The People's Republic of China was the first country to allow commercialized transgenic plants, introducing a virus-resistant tobacco in 1992, which was withdrawn in 1997. The first genetically modified crop approved for sale in the U.S., in 1994, was the *FlavrSavr* tomato. It had a longer shelf life, because it took longer to soften after ripening. In 1994, the European Union approved tobacco engineered to tolerate the herbicide bromoxynil, making it the first commercially genetically engineered crop marketed in Europe.

In 1995, Bt Potato was approved by the US Environmental Protection Agency, making it the country's first-pesticide producing crop. In 1995 canola with modified oil composition (Calgene), Bt maize (Ciba-Geigy), bromoxynil-tolerant cotton (Calgene), Bt cotton (Monsanto), glyphosate-tolerant soybeans (Monsanto), virus-tolerant squash (Asgrow), and additional delayed ripening tomatoes (DNAP, Zeneca/Peto and Monsanto) were approved. As of mid-1996, 35 approvals had been granted to commercially grow 8 transgenic crops and one flower crop (carnation), with 8 different traits in 6 countries plus the EU. In 2000, Vitamin A-enriched golden rice was developed, though as of 2016 it was not yet in commercial production.

In 2013 the leaders of the three research teams that first applied genetic engineering to crops, Robert Fraley, Marc Van Montagu and Mary-Dell Chilton, were awarded the World Food Prize for improving the "quality, quantity or availability" of food in the world.

In the US, by 2014, 94% of the planted area of soybeans, 96% of cotton and 93% of corn were genetically modified varieties. In developing countries, about 18 million farmers planted 54% of GM crops worldwide by 2013.

Methods

Genetically engineered crops have genes added or removed using genetic engineering techniques, originally including gene guns, electroporation, microinjection and agrobacterium. More recently, CRISPR and TALEN offered much more precise and convenient editing techniques.

Gene guns (also known as biolistics) "shoot" (direct high energy particles or radiations against) target genes into plant cells. It is the most common method. DNA is bound to tiny particles of gold or tungsten which are subsequently shot into plant tissue or single plant cells under high pressure. The accelerated particles penetrate both the cell wall and membranes. The DNA separates from the metal and is integrated into plant DNA inside the nucleus. This method has been applied successfully for many cultivated crops, especially monocots like wheat or maize, for which transformation using *Agrobacterium tumefaciens* has been less successful. The major disadvantage of this procedure is that serious damage can be done to the cellular tissue.

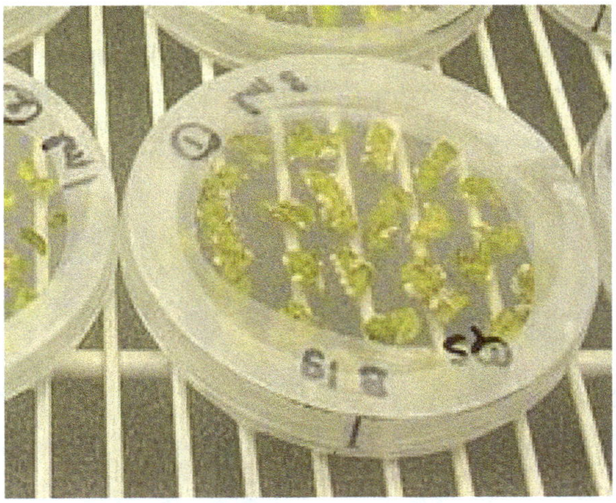

Plants (*Solanum chacoense*) being transformed using agrobacterium.

Agrobacterium tumefaciens-mediated transformation is another common technique. Agrobacteria are natural plant parasites. Their natural ability to transfer genes provides another engineering method. To create a suitable environment for themselves, these Agrobacteria insert their genes into plant hosts, resulting in a proliferation of modified plant cells near the soil level (crown gall). The genetic information for tumor growth is encoded on a mobile, circular DNA fragment (plasmid). When *Agrobacterium* infects a plant, it transfers this T-DNA to a random site in the plant genome. When used in genetic engineering the bacterial T-DNA is removed from the bacterial plasmid and replaced with the desired foreign gene. The bacterium is a vector, enabling transportation of foreign genes into plants. This method works especially well for dicotyledonous plants like potatoes, tomatoes, and tobacco. Agrobacteria infection is less successful in crops like wheat and maize.

Electroporation is used when the plant tissue does not contain cell walls. In this technique, "DNA enters the plant cells through miniature pores which are temporarily caused by electric pulses."

Microinjection is used to directly inject foreign DNA into cells. Plant scientists, backed by results of modern comprehensive profiling of crop composition, point out that crops modified using GM techniques are less likely to have unintended changes than are conventionally bred crops.

In research tobacco and *Arabidopsis thaliana* are the most frequently modified plants, due to well-developed transformation methods, easy propagation and well studied genomes. They serve as model organisms for other plant species.

Introducing new genes into plants requires a promoter specific to the area where the gene is to be expressed. For instance, to express a gene only in rice grains and not in leaves, an endosperm-specific promoter is used. The codons of the gene must be optimized for the organism due to codon usage bias.

Types of Modifications

Transgenic

Transgenic plants have genes inserted into them that are derived from another species. The inserted genes can come from species within the same kingdom (plant to plant), or between kingdoms (for example, bacteria to plant). In many cases the inserted DNA has to be modified slightly in order to be correctly and efficiently expressed in the host organism. Transgenic plants are used to express proteins, like the cry toxins from *B. thuringiensis*, herbicide-resistant genes, antibodies,and antigens for vaccinations. A study led by the European Food Safety Authority (EFSA) also found viral genes in transgenic plants.

Transgenic maize containing a gene from the bacteria *Bacillus thuringiensis*.

Transgenic carrots have been used to produce the drug Taliglucerase alfa which is used to treat Gaucher's disease. In the laboratory, transgenic plants have been modified to increase photosynthesis (currently about 2% at most plants versus the theoretic potential of 9–10%). This is possible by changing the rubisco enzyme (i.e. changing C_3 plants into C_4 plants), by placing the rubisco in a carboxysome, by adding CO_2 pumps in the cell wall, or by changing the leaf form or size.Plants have been engineered to exhibit bioluminescence that may become a sustainable alternative to electric lighting.

Cisgenic

Cisgenic plants are made using genes found within the same species or a closely related one, where conventional plant breeding can occur. Some breeders and scientists argue that cisgenic modification is useful for plants that are difficult to crossbreed by conventional means (such as potatoes), and that plants in the cisgenic category should not require the same regulatory scrutiny as transgenics.

Subgenic

Genetically modified plants can also be developed using gene knockdown or gene knockout to alter the genetic makeup of a plant without incorporating genes from other plants. In 2014, Chinese researcher Gao Caixia filed patents on the creation of a strain of wheat that is resistant to powdery mildew. The strain lacks genes that encode proteins that repress defenses against the mildew. The researchers deleted all three copies of the genes from wheat's hexaploid genome. Gao used the TALENs and CRISPR gene editing tools without adding or changing any other genes. No field trials were immediately planned. The CRISPR technique has also been used by Penn State researcher Yinong Yang to modify white button mushrooms (*Agaricus bisporus*) to be non-browning, and by DuPont Pioneer to make a new variety of corn.

Multiple Trait Integration

With multiple trait integration, several new traits may be integrated into a new crop.

Economics

GM food's economic value to farmers is one of its major benefits, including in developing nations. A 2010 study found that Bt corn provided economic benefits of $6.9 billion over the previous 14 years in five Midwestern states. The majority ($4.3 billion) accrued to farmers producing non-Bt corn. This was attributed to European corn borer populations reduced by exposure to Bt corn, leaving fewer to attack conventional corn nearby. Agriculture economists calculated that "world surplus [increased by] $240.3 million for 1996. Of this total, the largest share (59%) went to U.S. farmers. Seed company Monsanto received the next largest share (21%), followed by US consumers (9%), the rest of the world (6%), and the germplasm supplier, Delta & Pine Land Company of Mississippi (5%)."

According to the International Service for the Acquisition of Agri-biotech Applications (ISAAA), in 2014 approximately 18 million farmers grew biotech crops in 28 countries; about 94% of the farmers were resource-poor in developing countries. 53% of the global biotech crop area of 181.5 million hectares was grown in 20 developing countries. PG Economics comprehensive 2012 study concluded that GM crops increased farm incomes worldwide by $14 billion in 2010, with over half this total going to farmers in developing countries.

Critics challenged the claimed benefits to farmers over the prevalence of biased observers and by the absence of randomized controlled trials. The main Bt crop grown by small farmers in developing countries is cotton. A 2006 review of Bt cotton findings by agricultural economists concluded, "the overall balance sheet, though promising, is mixed. Economic returns are highly variable over years, farm type, and geographical location".

In 2013 the European Academies Science Advisory Council (EASAC) asked the EU to allow the development of agricultural GM technologies to enable more sustainable agriculture, by employing fewer land, water, and nutrient resources. EASAC also criticizes the EU's "time-consuming and expensive regulatory framework" and said that the EU had fallen behind in the adoption of GM technologies.

Participants in agriculture business markets include seed companies, agrochemical companies, distributors, farmers, grain elevators and universities that develop new crops/traits and whose agricultural extensions advise farmers on best practices. According to a 2012 review based on data from the late 1990s and early 2000s, much of the GM crop grown each year is used for livestock feed and increased demand for meat leads to increased demand for GM feed crops. Feed grain usage as a percentage of total crop production is 70% for corn and more than 90% of oil seed meals such as soybeans. About 65 million metric tons of GM corn grains and about 70 million metric tons of soybean meals derived from GM soybean become feed.

In 2014 the global value of biotech seed was US$15.7 billion; US$11.3 billion (72%) was in industrial countries and US$4.4 billion (28%) was in the developing countries. In 2009, Monsanto had $7.3 billion in sales of seeds and from licensing its technology; DuPont, through its Pioneer subsidiary, was the next biggest company in that market. As of 2009, the overall

Roundup line of products including the GM seeds represented about 50% of Monsanto's business.

Some patents on GM traits have expired, allowing the legal development of generic strains that include these traits. For example, generic glyphosate-tolerant GM soybean is now available. Another impact is that traits developed by one vendor can be added to another vendor's proprietary strains, potentially increasing product choice and competition. The patent on the first type of *Roundup Ready* crop that Monsanto produced (soybeans) expired in 2014 and the first harvest of off-patent soybeans occurs in the spring of 2015.Monsanto has broadly licensed the patent to other seed companies that include the glyphosate resistance trait in their seed products. About 150 companies have licensed the technology, including Syngenta and DuPont Pioneer.

Yield

In 2014, the largest review yet concluded that GM crops' effects on farming were positive. The meta-analysisconsidered all published English-language examinations of the agronomic and economic impacts between 1995 and March 2014 for three major GM crops: soybean, maize, and cotton. The study found that herbicide-tolerant crops have lower production costs, while for insect-resistant crops the reduced pesticide use was offset by higher seed prices, leaving overall production costs about the same.

Yields increased 9% for herbicide tolerance and 25% for insect resistant varieties. Farmers who adopted GM crops made 69% higher profits than those who did not. The review found that GM crops help farmers in developing countries, increasing yields by 14 percentage points.

The researchers considered some studies that were not peer-reviewed and a few that did not report sample sizes. They attempted to correct for publication bias, by considering sources beyond academic journals. The large data set allowed the study to control for potentially confounding variables such as fertilizer use. Separately, they concluded that the funding source did not influence study results.

Traits

GM crops grown today, or under development, have been modified with various traits. These traits include improved shelf life, disease resistance, stress resistance, herbicide resistance, pest resistance, production of useful goods such as biofuel or drugs, and ability to absorb toxins and for use in bioremediation of pollution.

Recently, research and development has been targeted to enhancement of crops that are locally important in developing countries, such as insect-resistant cowpea for Africa and insect-resistant brinjal (eggplant).

Extended Shelf Life

The first genetically modified crop approved for sale in the U.S. was the *FlavrSavr* tomato, which had a longer shelf life. First sold in 1994, FlavrSavr tomato production ceased in 1997. It is no longer on the market. In 2014, the USDA approved a GM potato that prevents bruising.

In February 2015 Arctic Apples were approved by the USDA, becoming the first genetically modified apple approved for US sale. Gene silencing was used to reduce the expression of polyphenol oxidase (PPO), thus preventing enzymatic browning of the fruit after it has been sliced open. The trait was added to Granny Smithand Golden Delicious varieties. The trait includes a bacterial antibiotic resistance gene that provides resistance to the antibiotic kanamycin. The genetic engineering involved cultivation in the presence of kanamycin, which allowed only resistant cultivars to survive. Humans consuming apples do not acquire kanamycin resistance, per arcticapple.com. The FDA approved the apples in March 2015.

Improved Photosynthesis

Plants use non-photochemical quenching (NPQ) to protect them from excessive amounts of sunlight. Plants can switch on the quenching mechanism almost instantaneously but it takes much longer for it to switch off again. During the time that it is switched off, the amount of energy that is wasted increases. A genetic modification in three genes allows to correct this (in a trial with tobacco plants). As a result, yields were 14-20% higher, in terms of the weight of the dry leaves harvested. The plants had bigger leaves, were taller and had more vigorous roots.

Another improvement that can be made on the photosynthesis process (with C3 pathway plants) is on photorespiration. By inserting the C4 pathway into C3 plants, productivity may increase by as much as 50% for cereal crops, such as rice.

Improved Nutritional Value

Edible Oils

Some GM soybeans offer improved oil profiles for processing. Camelina sativa has been modified to produce plants that accumulate high levels of oils similar to fish oils.

Vitamin Enrichment

Golden rice, developed by the International Rice Research Institute (IRRI), provides greater amounts of vitamin A targeted at reducing vitamin A deficiency. As of January 2016, golden rice has not yet been grown commercially in any country.

Toxin Reduction

A genetically modified cassava under development offers lower cyanogen glucosides and enhanced protein and other nutrients (called BioCassava).

In November 2014, the USDA approved a potato that prevents bruising and produces less acrylamide when fried. They do not employ genes from non-potato species. The trait was added to the Russet Burbank, Ranger Russet and Atlantic varieties.

Stress Resistance

Plants have been engineered to tolerate non-biological stressors, such as drought, frost,and high soil salinity. In 2011, Monsanto's DroughtGard maize became the first drought-resistant GM crop to receive US marketing approval.

Drought resistance occurs by modifying the plant's genes responsible for the mechanism known as the crassulacean acid metabolism (CAM), which allows the plants to survive despite low water levels. This holds promise for water-heavy crops such as rice, wheat, soybeans and poplar to accelerate their adaptation to water-limited environments. Several salinity tolerance mechanisms have been identified in salt-tolerant crops. For example, rice, canola and tomato crops have been genetically modified to increase their tolerance to salt stress.

Herbicides

Glyphosate

As of 1999, the most prevalent GM trait was glyphosate-tolerance. Glyphosate (the active ingredient in Roundup and other herbicide products) kills plants by interfering with the shikimate pathway in plants, which is essential for the synthesis of the aromatic amino acids phenylalanine, tyrosine, and tryptophan. The shikimate pathway is not present in animals, which instead obtain aromatic amino acids from their diet. More specifically, glyphosate inhibits the enzyme 5-enolpyruvylshikimate-3-phosphate synthase (EPSPS).

This trait was developed because the herbicides used on grain and grass crops at the time were highly toxic and not effective against narrow-leaved weeds. Thus, developing crops that could withstand spraying with glyphosate would both reduce environmental and health risks, and give an agricultural edge to the farmer.

Some micro-organisms have a version of EPSPS that is resistant to glyphosate inhibition. One of these was isolated from an *Agrobacterium* strain CP4 (CP4 EPSPS) that was resistant to glyphosate. The CP4 EPSPS gene was engineered for plant expression by fusing the 5' end of the gene to a chloroplast transit peptide derived from the petunia EPSPS. This transit peptide was used because it had shown previously an ability to deliver bacterial EPSPS to the chloroplasts of other plants. This CP4 EPSPS gene was cloned and transfected into soybeans.

The plasmid used to move the gene into soybeans was PV-GMGTO4. It contained three bacterial genes, two CP4 EPSPS genes, and a gene encoding beta-glucuronidase (GUS) from *Escherichia coli* as a marker. The DNA was injected into the soybeans using the particle acceleration method. Soybean cultivar A54O3 was used for the transformation.

Bromoxynil

Tobacco plants have been engineered to be resistant to the herbicide bromoxynil.

Glufosinate

Crops have been commercialized that are resistant to the herbicide glufosinate, as well. Crops engineered for resistance to multiple herbicides to allow farmers to use a mixed group of two, three, or four different chemicals are under development to combat growing herbicide resistance.

2,4-D

In October 2014 the US EPA registered Dow's Enlist Duo maize, which is genetically modified to be resistant to both glyphosate and 2,4-D, in six states. Inserting a bacterial aryloxyalkanoate dioxygenase gene, *aad1*makes the corn resistant to 2,4-D. The USDA had approved maize and soybeans with the mutation in September 2014.

Dicamba

Monsanto has requested approval for a stacked strain that is tolerant of both glyphosate and dicamba. The request includes plans for avoiding herbicide drift to other crops. Significant damage to other non-resistant crops occurred from dicamba formulations intended to reduce volatilization drifting when sprayed on resistant soybeans in 2017. The newer dicamba formulation labels specify to not spray when average wind speeds are above 10–15 miles per hour (16–24 km/h) to avoid particle drift, average wind speeds below 3 miles per hour (4.8 km/h) to avoid temperature inversions, and rain or high temperatures are in the next day forecast. However, these conditions typically only occur during June and July for a few hours at a time.

Pest Resistance

Insects

Tobacco, corn, rice and some other crops have been engineered to express genes encoding for insecticidal proteins from Bacillus thuringiensis (Bt). The introduction of Bt crops during the period between 1996 and 2005 has been estimated to have reduced the total volume of insecticide active ingredient use in the United States by over 100 thousand tons. This represents a 19.4% reduction in insecticide use.

In the late 1990s, a genetically modified potato that was resistant to the Colorado potato beetle was withdrawn because major buyers rejected it, fearing consumer opposition.

Viruses

Papaya, potatoes, and squash have been engineered to resist viral pathogens such as cucumber mosaic viruswhich, despite its name, infects a wide variety of plants. Virus resistant papaya were developed in response to a papaya ringspot virus (PRV) outbreak in Hawaii in the late 1990s. They incorporate PRV DNA. By 2010, 80% of Hawaiian papaya plants were genetically modified.

Potatoes were engineered for resistance to potato leaf roll virus and Potato virus Y in 1998. Poor sales led to their market withdrawal after three years.

Yellow squash that were resistant to at first two, then three viruses were developed, beginning in the 1990s. The viruses are watermelon, cucumber and zucchini/courgette yellow mosaic. Squash was the second GM crop to be approved by US regulators. The trait was later added to zucchini.

Many strains of corn have been developed in recent years to combat the spread of Maize dwarf mosaic virus, a costly virus that causes stunted growth which is carried in Johnson grass and spread by aphid insect vectors. These strands are commercially available although the resistance is not standard among GM corn variants.

By-products

Drugs

In 2012, the FDA approved the first plant-produced pharmaceutical, a treatment for Gaucher's Disease.Tobacco plants have been modified to produce therapeutic antibodies.

Biofuel

Algae is under development for use in biofuels. Researchers in Singapore were working on GM jatropha for biofuel production. Syngenta has USDA approval to market a maize trademarked Enogen that has been genetically modified to convert its starch to sugar for ethanol. Some trees have been genetically modified to either have less lignin, or to express lignin with chemically labile bonds. Lignin is the critical limiting factor when using wood to make bio-ethanol because lignin limits the accessibility of cellulose microfibrils to depolymerization by enzymes. Besides with trees, the chemically labile lignin bonds are also very useful for cereal crops such as maize, barley, and oats.

Materials

Companies and labs are working on plants that can be used to make bioplastics. Potatoes that produce industrially useful starches have been developed as well. Oilseed can be modified to produce fatty acids for detergents, substitute fuels and petrochemicals.

Bioremediation

Scientists at the University of York developed a weed (*Arabidopsis thaliana*) that contains genes from bacteria that could clean TNT and RDX-explosive soil contaminants in 2011. 16 million hectares in the US (1.5% of the total surface) are estimated to be contaminated with TNT and RDX. However *A. thaliana* was not tough enough for use on military test grounds. Modifications in 2016 included switchgrass and bentgrass.

Genetically modified plants have been used for bioremediation of contaminated soils. Mercury, selenium and organic pollutants such as polychlorinated biphenyls (PCBs).

Marine environments are especially vulnerable since pollution such as oil spills are not containable. In addition to anthropogenic pollution, millions of tons of petroleum annually enter the marine environment from natural seepages. Despite its toxicity, a considerable fraction of petroleum oil entering marine systems is eliminated by the hydrocarbon-degrading activities of microbial communities. Particularly successful is a recently discovered group of specialists, the so-called hydrocarbonoclastic bacteria (HCCB) that may offer useful genes.

Asexual Reproduction

Crops such as maize reproduce sexually each year. This randomizes which genes get propagated to the next generation, meaning that desirable traits can be lost. To maintain a high-quality crop, some farmers purchase seeds every year. Typically, the seed company maintains two inbred varieties and crosses them into a hybridstrain that is then sold. Related plants like sorghum and gamma grass are able to perform apomixis, a form of asexual reproduction that keeps the plant's DNA intact. This trait is apparently controlled by a single dominant gene, but traditional breeding has been unsuccessful in creating asexually-reproducing maize. Genetic engineering offers another route to this goal. Successful modification would allow farmers to replant harvested seeds that retain desirable traits, rather than relying on purchased seed.

Other

Genetic modifications to some crops also exist, which make it easier to process the crop. Also, some crops (such as tomatoes) have been genetic modified to contain no seed at all.

Farming Practices

Bacillus Thuringiensis

Constant exposure to a toxin creates evolutionary pressure for pests resistant to that toxin. Overreliance on glyphosate and a reduction in the diversity of weed management practices allowed the spread of glyphosate resistance in 14 weed species in the US, and in soybeans.

To reduce resistance to *Bacillus thuringiensis* (Bt) crops, the 1996 commercialization of transgenic cotton and maize came with a management strategy to prevent insects from becoming resistant. Insect resistance management plans are mandatory for Bt crops. The aim is to encourage a large population of pests so that any (recessive) resistance genes are diluted within the population. Resistance lowers evolutionary fitness in the absence of the stressor, Bt. In refuges, non-resistant strains outcompete resistant ones.

With sufficiently high levels of transgene expression, nearly all of the heterozygotes (S/s), i.e., the largest segment of the pest population carrying a resistance allele, will be killed before maturation, thus preventing transmission of the resistance gene to their progeny. Refuges (i. e., fields of non-transgenic plants) adjacent to transgenic fields increases the likelihood that homozygous resistant (s/s) individuals and any surviving heterozygotes will mate with susceptible (S/S) individuals from the refuge, instead of with other individuals carrying the resistance allele. As a result, the resistance gene frequency in the population remains lower.

Complicating factors can affect the success of the high-dose/refuge strategy. For example, if the temperature is not ideal, thermal stress can lower Bt toxin production and leave the plant more susceptible. More importantly, reduced late-season expression has been documented, possibly resulting from DNA methylation of the promoter. The success of the high-dose/refuge strategy has successfully maintained the value of Bt crops. This success has depended on factors independent of management strategy, including low initial resistance allele frequencies, fitness costs associated with resistance, and the abundance of non-Bt host plants outside the refuges.

Companies that produce Bt seed are introducing strains with multiple Bt proteins. Monsanto did this with Bt cotton in India, where the product was rapidly adopted. Monsanto has also; in an attempt to simplify the process of implementing refuges in fields to comply with Insect Resistance

Management(IRM) policies and prevent irresponsible planting practices; begun marketing seed bags with a set proportion of refuge (non-transgenic) seeds mixed in with the Bt seeds being sold. Coined "Refuge-In-a-Bag" (RIB), this practice is intended to increase farmer compliance with refuge requirements and reduce additional labor needed at planting from having separate Bt and refuge seed bags on hand. This strategy is likely to reduce the likelihood of Bt-resistance occurring for corn rootworm, but may increase the risk of resistance for lepidopteran corn pests, such as European corn borer. Increased concerns for resistance with seed mixtures include partially resistant larvae on a Bt plant being able to move to a susceptible plant to survive or cross pollination of refuge pollen on to Bt plants that can lower the amount of Bt expressed in kernels for ear feeding insects.

Herbicide Resistance

Best management practices (BMPs) to control weeds may help delay resistance. BMPs include applying multiple herbicides with different modes of action, rotating crops, planting weed-free seed, scouting fields routinely, cleaning equipment to reduce the transmission of weeds to other fields, and maintaining field borders. The most widely planted GM crops are designed to tolerate herbicides. By 2006 some weed populations had evolved to tolerate some of the same herbicides. Palmer amaranth is a weed that competes with cotton. A native of the southwestern US, it traveled east and was first found resistant to glyphosate in 2006, less than 10 years after GM cotton was introduced.

Plant Protection

Farmers generally use less insecticide when they plant Bt-resistant crops. Insecticide use on corn farms declined from 0.21 pound per planted acre in 1995 to 0.02 pound in 2010. This is consistent with the decline in European corn borer populations as a direct result of Bt corn and cotton. The establishment of minimum refuge requirements helped delay the evolution of Bt resistance. However, resistance appears to be developing to some Bt traits in some areas.

Tillage

By leaving at least 30% of crop residue on the soil surface from harvest through planting, conservation tillagereduces soil erosion from wind and water, increases water retention, and reduces soil degradation as well as water and chemical runoff. In addition, conservation tillage reduces the carbon footprint of agriculture. A 2014 review covering 12 states from 1996 to 2006, found that a 1% increase in herbicde-tolerant (HT) soybean adoption leads to a 0.21% increase in conservation tillage and a 0.3% decrease in quality-adjusted herbicide use.

Regulation

The regulation of genetic engineering concerns the approaches taken by governments to assess and manage the risks associated with the development and release of genetically modified crops. There are differences in the regulation of GM crops between countries, with some of the most marked differences occurring between the US and Europe. Regulation varies in a given country depending on the intended use of each product. For example, a crop not intended for food use is generally not reviewed by authorities responsible for food safety.

Production

In 2013, GM crops were planted in 27 countries; 19 were developing countries and 8 were developed countries. 2013 was the second year in which developing countries grew a majority (54%) of the total GM harvest. 18 million farmers grew GM crops; around 90% were small-holding farmers in developing countries.

Country	2013– GM planted area (million hectares)	Biotech crops
US	70.1	Maize, Soybean, Cotton, Canola, Sugarbeet, Alfalfa, Papaya, Squash
Brazil	40.3	Soybean, Maize, Cotton
Argentina	24.4	Soybean, Maize, Cotton
India	11.0	Cotton
Canada	10.8	Canola, Maize, Soybean, Sugarbeet
Total	175.2	----

The United States Department of Agriculture (USDA) reports every year on the total area of GM crop varieties planted in the United States. According to National Agricultural Statistics Service, the states published in these tables represent 81–86 percent of all corn planted area, 88–90 percent of all soybean planted area, and 81–93 percent of all upland cotton planted area (depending on the year).

Global estimates are produced by the International Service for the Acquisition of Agri-biotech Applications(ISAAA) and can be found in their annual reports, “Global Status of Commercialized Transgenic Crops”.

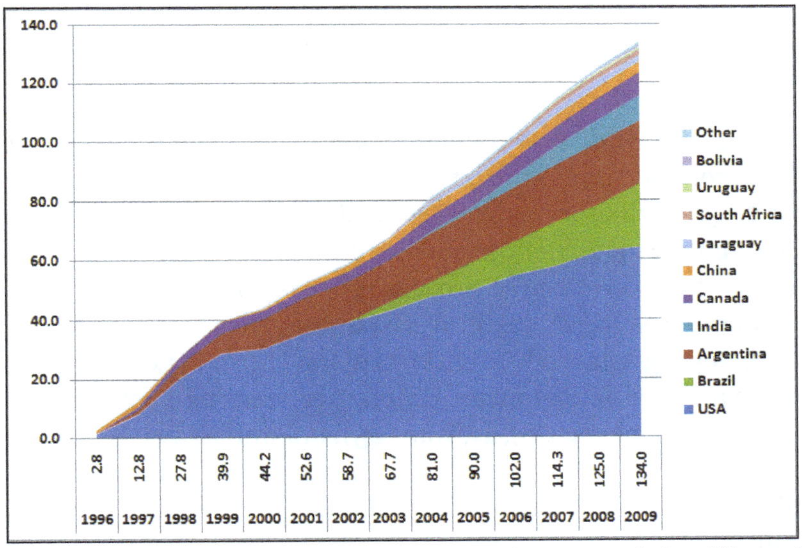

Land area used for genetically modified crops by country (1996–2009), in millions of hectares. In 2011, the land area used was 160 million hectares, or 1.6 million square kilometers.

Farmers have widely adopted GM technology. Between 1996 and 2013, the total surface area of land cultivated with GM crops increased by a factor of 100, from 17,000 square kilometers (4,200,000 acres) to 1,750,000 km^2 (432 million acres). 10% of the world’s arable land was planted with GM crops in 2010. As of 2011, 11 different transgenic crops were grown commercially on 395 million acres (160 million hectares) in 29 countries such as the US, Brazil, Argentina, India, Canada, China, Paraguay, Pakistan, South Africa, Uruguay, Bolivia, Australia, Philippines, Myanmar, Burkina Faso, Mexico and Spain. One of the key reasons for this widespread adoption is the perceived economic benefit the technology brings to farmers. For example, the system of planting glyphosate-resistant seed and then applying glyphosate once plants emerged provided farmers with the opportunity to dramatically increase the yield from a given plot of land, since this allowed them to plant rows closer together. Without it, farmers had to plant rows far enough apart to control post-emergent weeds with mechanical tillage. Likewise, using Bt seeds means that farmers do not have to purchase insecticides, and then invest time, fuel, and equipment in applying them. However critics have disputed whether yields are higher and whether chemical use is less, with GM crops.

In the US, by 2014, 94% of the planted area of soybeans, 96% of cotton and 93% of corn were genetically modified varieties. Genetically modified soybeans carried herbicide-tolerant traits only, but maize and cotton carried both herbicide tolerance and insect protection traits (the latter largely Bt protein). These constitute "input-traits" that are aimed to financially benefit the producers, but may have indirect environmental benefits and cost benefits to consumers. The Grocery Manufacturers of America estimated in 2003 that 70–75% of all processed foods in the U.S. contained a GM ingredient.

Europe grows relatively few genetically engineered crops with the exception of Spain, where one fifth of maize is genetically engineered, and smaller amounts in five other countries. The EU had a 'de facto' ban on the approval of new GM crops, from 1999 until 2004. GM crops are now regulated by the EU. In 2015, genetically engineered crops are banned in 38 countries worldwide, 19 of them in Europe. Developing countries grew 54 percent of genetically engineered crops in 2013.

In recent years GM crops expanded rapidly in developing countries. In 2013 approximately 18 million farmers grew 54% of worldwide GM crops in developing countries. 2013's largest increase was in Brazil (403,000 km^2versus 368,000 km^2 in 2012). GM cotton began growing in India in 2002, reaching 110,000 km^2 in 2013.

According to the 2013 ISAAA brief: "a total of 36 countries (35 + EU-28) have granted regulatory approvals for biotech crops for food and/or feed use and for environmental release or planting since 1994 a total of 2,833 regulatory approvals involving 27 GM crops and 336 GM events (NB: an "event" is a specific genetic modification in a specific species) have been issued by authorities, of which 1,321 are for food use (direct use or processing), 918 for feed use (direct use or processing) and 599 for environmental release or planting. Japan has the largest number (198), followed by the U.S.A. (165, not including "stacked" events), Canada (146), Mexico (131), South Korea (103), Australia (93), New Zealand (83), European Union (71 including approvals that have expired or under renewal process), Philippines (68), Taiwan (65), Colombia (59), China (55) and South Africa (52). Maize has the largest number (130 events in 27 countries), followed by cotton (49 events in 22 countries), potato (31 events in 10 countries), canola (30 events in 12 countries) and soybean (27 events in 26 countries).

References

- Genetic-engineering, science: britannica.com, Retrieved 10 July, 2019
- Chen I, Dubnau D (March 2004). "DNA uptake during bacterial transformation". Nature Reviews. Microbiology. 2 (3): 241–9. Doi:10.1038/nrmicro844. PMID 15083159

- Genetically-modified-organism, science: britannica.com, Retrieved 13 March, 2019
- Nicholl DS (2008-05-29). An Introduction to Genetic Engineering. Cambridge University Press. P. 34. ISBN 9781139471787
- ISAAA 2013 Annual Report Executive Summary, Global Status of Commercialized Biotech/GM Crops: 2013 ISAAA Brief 46-2013, Retrieved 6 August 2014
- Setlow JK (2002-10-31). Genetic Engineering: Principles and Methods. Springer Science & Business Media. P. 109. ISBN 9780306472800
- Lynch, Diahanna; Vogel, David (5 April 2001). "The Regulation of gmos in Europe and the United States: A Case-Study of Contemporary European Regulatory Politics". Council on Foreign Relations. Retrieved 24 February 2016

Heredity and Evolution

CHAPTER 5

The passing on of traits from parents to their offspring through sexual or asexual reproduction is known as heredity. The change in the heritable characteristics of biological populations over successive generations is known as evolution. The chapter closely examines the key concepts of heredity and evolution such as Mendelian inheritance, speciation and common descent to provide an extensive understanding of the subject.

HEREDITY

Heredity is the sum of all biological processes by which particular characteristics are transmitted from parents to their offspring. The concept of heredity encompasses two seemingly paradoxical observations about organisms: the constancy of a species from generation to generation and the variation among individuals within a species. Constancy and variation are actually two sides of the same coin, as becomes clear in the study of genetics. Both aspects of heredity can be explained by genes, the functional units of heritable material that are found within all living cells. Every member of a species has a set of genes specific to that species. It is this set of genes that provides the constancy of the species. Among individuals within a species, however, variations can occur in the form each gene takes, providing the genetic basis for the fact that no two individuals (except identical twins) have exactly the same traits.

The set of genes that an offspring inherits from both parents, a combination of the genetic material of each, is called the organism's genotype. The genotype is contrasted to the phenotype, which is the organism's outward appearance and the developmental outcome of its genes. The phenotype includes an organism's bodily structures, physiological processes, and behaviors. Although the genotype determines the broad limits of the features an organism can develop, the features that actually develop, i.e., the phenotype, depend on complex interactions between genes and their environment. The genotype remains constant throughout an organism's lifetime; however, because the organism's internal and external environments change continuously, so does its phenotype. In conducting genetic studies, it is crucial to discover the degree to which the observable trait is attributable to the pattern of genes in the cells and to what extent it arises from environmental influence.

Because genes are integral to the explanation of hereditary observations, genetics also can be defined as the study of genes. Discoveries into the nature of genes have shown that genes are important determinants of all aspects of an organism's makeup. For this reason, most areas of biological research now have a genetic component, and the study of genetics has a position of central importance in biology. Genetic research also has demonstrated that virtually all organisms on this planet have similar genetic systems, with genes that are built on the same chemical principle and that function according to similar mechanisms. Although species differ in the sets of genes they contain, many similar genes are found across a wide range of species. For example, a large proportion of genes in baker's yeast are also present in humans. This similarity in genetic makeup between organisms that have such disparate phenotypes can be explained by the evolutionary relatedness of virtually all life-forms on Earth. This genetic unity has radically reshaped the understanding of the relationship between humans and all other organisms. Genetics also has had a profound impact on human affairs. Throughout history humans have created or improved many different medicines, foods, and textiles by subjecting plants, animals, and microbes to the ancient techniques of selective breeding and to the modern methods of recombinant DNA technology. In recent years medical researchers have begun to discover the role that genes play in disease. The significance of genetics only promises to become greater as the structure and function of more and more human genes are characterized.

Basic Features of Heredity

Prescientific Conceptions of Heredity

Heredity was for a long time one of the most puzzling and mysterious phenomena of nature. This was so because the sex cells, which form the bridge across which heredity must pass between the generations, are usually invisible to the naked eye. Only after the invention of the microscope early in the 17th century and the subsequent discovery of the sex cells could the essentials of heredity be grasped. Before that time, ancient Greek philosopher and scientist Aristotle (4th century BC) speculated that the relative contributions of the female and the male parents were very unequal; the female was thought to supply what he called the "matter" and the male the "motion." The Institutes of Manu, composed in India between 100 and 300 AD, consider the role of the female like that of the field and of the male like that of the seed; new bodies are formed "by the united operation of the seed and the field." In reality both parents transmit the heredity pattern equally, and, on average, children resemble their mothers as much as they do their fathers. Nevertheless, the female and male sex cells may be very different in size and structure; the mass of an egg cell is sometimes millions of times greater than that of a spermatozoon.

The ancient Babylonians knew that pollen from a male date palm tree must be applied to the pistils of a female tree to produce fruit. German botanist Rudolph Jacob Camerarius showed in 1694 that the same is true in corn (maize). Swedish botanist and explorer Carolus Linnaeus in 1760 and German botanist Josef Gottlieb Kölreuter, in a series of works published from 1761 to 1798, described crosses of varieties and species of plants. They found that these hybrids were, on the whole, intermediate between the parents, although in some characteristics they might be closer to one parent and in others closer to the other parent. Kölreuter compared the offspring of reciprocal crosses—i.e., of crosses of variety A functioning as a female to variety B as a male and the reverse, variety B as a female to A as a male. The hybrid progenies of these reciprocal crosses were usually alike, indicating that, contrary to the belief of Aristotle, the hereditary endowment of the progeny was derived equally from the female and the male parents. Many more experiments on plant hybrids were made in the 1800s. These investigations also revealed that hybrids were usually intermediate between the parents. They incidentally recorded most of the facts that later led Gregor Mendel to formulate his celebrated rules and to found the theory of the gene. Apparently, none of Mendel's predecessors saw the significance of the data that were being accumulated. The general intermediacy of hybrids seemed to agree best with the belief that heredity was transmitted from parents to offspring by "blood," and this belief was accepted by most 19th-century biologists, including English naturalist Charles Darwin.

The blood theory of heredity, if this notion can be dignified with such a name, is really a part of the folklore antedating scientific biology. It is implicit in such popular phrases as "half blood," "new blood," and "blue blood." It does not mean that heredity is actually transmitted through the red liquid in blood vessels; the essential point is the belief that a parent transmits to each child all its characteristics and that the hereditary endowment of a child is an alloy, a blend of the endowments of its parents, grandparents, and more-remote ancestors. This idea appeals to those who pride themselves on having a noble or remarkable "blood" line. It strikes a snag, however, when one observes that a child has some characteristics that are not present in either parent but are present in some other relatives or were present in more-remote ancestors. Even more often, one sees that brothers and sisters, though showing a family resemblance in some traits, are clearly different in others. How could the same parents transmit different "bloods" to each of their children?

Mendel disproved the blood theory. He showed (1) that heredity is transmitted through factors (now called genes) that do not blend but segregate, (2) that parents transmit only one-half of the genes they have to each child, and they transmit different sets of genes to different children, and (3) that, although brothers and sisters receive their heredities from the same parents, they do not receive the same heredities (an exception is identical twins). Mendel thus showed that, even if the eminence of some ancestor were entirely the reflection of his genes, it is quite likely that some of his descendants, especially the more remote ones, would not inherit these "good" genes at all. In sexually reproducing organisms, humans included, every individual has a unique hereditary endowment.

Lamarckism—a school of thought named for the 19th-century pioneer French biologist and evolutionist Jean-Baptiste de Monet, chevalier de Lamarck—assumed that characters acquired during an individual's life are inherited by his progeny, or, to put it in modern terms, that the modifications wrought by the environment in the phenotype are reflected in similar changes in the genotype. If this were so, the results of physical exercise would make exercise much easier or even dispensable in a person's offspring. Not only Lamarck but also other 19th-century biologists, including Darwin, accepted the inheritance of acquired traits. It was questioned by German biologist August Weismann, whose famous experiments in the late 1890s on the amputation of tails in generations of mice showed that such modification resulted neither in disappearance nor even in shortening of the tails of the descendants. Weismann concluded that the hereditary endowment of the organism, which he called the germ plasm, is wholly separate and is protected against the influences emanating from the rest of the body, called the somatoplasm, or soma. The germ plasm–somatoplasm are related to the genotype–phenotype concepts, but they are not identical and should not be confused with them.

The noninheritance of acquired traits does not mean that the genes cannot be changed by environmental influences; X-rays and other mutagens certainly do change them, and the genotype of a population can be altered by selection. It simply means that what is acquired by parents in their physique and intellect is not inherited by their children. Related to these misconceptions are the beliefs in "prepotency"—i.e., that some individuals impress their heredities on their progenies more effectively than others—and in "prenatal influences" or "maternal impressions"—i.e., that the events experienced by a pregnant female are reflected in the constitution of the child to be born. How ancient these beliefs are is suggested in the Book of Genesis, in which Laban produced spotted or striped progeny in sheep by showing the pregnant ewes striped hazel rods. Another such belief is "telegony," which goes back to Aristotle; it alleged that the heredity of an individual is influenced not only by his father but also by males with whom the female may have mated and who have caused previous pregnancies. Even Darwin, as late as 1868, seriously discussed an alleged case of telegony: that of a mare mated to a zebra and subsequently to an Arabian stallion, by whom the mare produced a foal with faint stripes on his legs. The simple explanation for this result is that such stripes occur naturally in some breeds of horses.

All these beliefs, from inheritance of acquired traits to telegony, must now be classed as superstitions. They do not stand up under experimental investigation and are incompatible with what is known about the mechanisms of heredity and about the remarkable and predictable properties of genetic materials. Nevertheless, some people still cling to these beliefs. Some animal breeders take telegony seriously and do not regard as purebred the individuals whose parents are admittedly "pure" but whose mothers had mated with males of other breeds. Soviet biologist and agronomist Trofim Denisovich Lysenko was able for close to a quarter of a century, roughly between 1938 and 1963, to make his special brand of Lamarckism the official creed in the Soviet Union and to suppress most of the teaching and research in orthodox genetics. He and his partisans published hundreds of articles and books allegedly proving their contentions, which effectively deny the achievements of biology for at least the preceding century. The Lysenkoists were officially discredited in 1964.

Mendelian Genetics

Discovery and Rediscovery of Mendel's Laws

Gregor Mendel published his work in the proceedings of the local society of naturalists in Brünn, Austria (now Brno, Czech Republic), in 1866, but none of his contemporaries appreciated its significance. It was not until 1900, 16 years after Mendel's death, that his work was rediscovered independently by botanists Hugo de Vries in Holland, Carl Erich Correns in Germany, and Erich Tschermak von Seysenegg in Austria. Like several investigators before him, Mendel experimented on hybrids of different varieties of a plant; he focused on the common pea plant (Pisum sativum). His methods differed in two essential respects from those of his predecessors. First, instead of trying to describe the appearance of whole plants with all their characteristics, Mendel followed the inheritance of single, easily visible and distinguishable traits, such as round versus wrinkled seed, yellow versus green seed, purple versus white flowers, and so on. Second, he made exact counts of the numbers of plants bearing each trait; it was from such quantitative data that he deduced the rules governing inheritance.

Since pea plants reproduce usually by self-pollination of their flowers, the varieties Mendel obtained from seedsmen were "pure"—i.e., descended for several to many generations from plants with similar traits. Mendel crossed them by deliberately transferring the pollen of one variety to the pistils of another; the resulting first-generation hybrids, denoted by the symbol F1, usually showed the traits of only one parent. For example, the crossing of yellow-seeded plants with green-seeded ones gave yellow seeds, and the crossing of purple-flowered plants with white-flowered ones gave purple-flowered plants. Traits such as the yellow-seed colour and the purple-flower colour Mendel called dominant; the green-seed colour and the white-flower colour he called recessive. It looked as if the yellow and purple "bloods" overcame or consumed the green and white "bloods."

That this was not so became evident when Mendel allowed the F1 hybrid plants to self-pollinate and produce the second hybrid generation, F2. Here, both the dominant and the recessive traits reappeared, as pure and uncontaminated as they were in the original parents (generation P). Moreover, these traits now appeared in constant proportions: about 3/4 of the plants in the second generation showed the dominant trait and 1/4 showed the recessive, a 3 to 1 ratio. It can be seen in the table that Mendel's actual counts were as close to the ideal ratio as one could expect, allowing for the sampling deviations present in all statistical data.

Pea plants with dominant and recessive characters obtained by Mendel in the second generation of hybrids.

Number dominant		Number recessive		Ratio
Round seed	5,474	Wrinkled seed	1,850	2.96:1
Yellow seed	6,022	Green seed	2,001	3.01:1
Purple flowers	705	White flowers	224	3.15:1
Tall plants	787	Short plants	277	2.84:1

Mendel concluded that the sex cells, the gametes, of the purple-flowered plants carried some factor that caused the progeny to develop purple flowers, and the gametes of the white-flowered variety had a variant factor that induced the development of white flowers. In 1909 the Danish biologist Wilhelm Ludvig Johannsen proposed to call these factors genes.

An example of one of Mendel's experiments will illustrate how the genes are transmitted and in what particular ratios. Let R stand for the gene for purple flowers and r for the gene for white flowers (dominant genes are conventionally symbolized by capital letters and recessive genes by lowercase letters). Since each pea plant contains a gene endowment half of whose set is derived from the mother and half from the father, each plant has two genes for flower colour. If the two genes are alike—for instance, both having come from white-flowered parents (rr)—the plant is termed a homozygote. The union of gametes with different genes gives a hybrid plant, termed a heterozygote (Rr). Since the gene R, for purple, is dominant over r, for white, the F_1 generation hybrids will show purple flowers. They are phenotypically purple, but their genotype contains both R and r genes, and these alternative (allelic or allelomorphic) genes do not blend or contaminate each other. Mendel inferred that, when a heterozygote forms its sex cells, the allelic genes segregate and pass to different gametes. This is expressed in the first law of Mendel, the law of segregation of unit genes. Equal numbers of gametes, ovules, or pollen grains are formed that contain the genes R and r. Now, if the gametes unite at random, then the F_2 generation should contain about $^1/_4$ white-flowered and $^3/_4$ purple-flowered plants. The white-flowered plants, which must be recessive homozygotes, bear the genotype rr. About $^1/_3$ of the plants exhibiting the dominant trait of purple flowers must be homozygotes, RR, and $^2/_3$ heterozygotes,

Rr. The prediction is tested by obtaining a third generation, F_3, from the purple-flowered plants; though phenotypically all purple-flowered, $^2/_3$ of this group of plants reveal the presence of the recessive gene allele, r, in their genotype by producing about $^1/_4$ white-flowered plants in the F_3 generation.

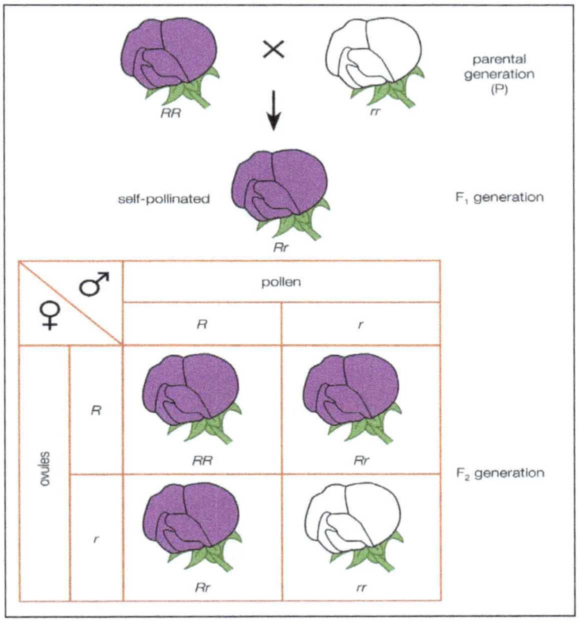

Mendel's law of segregationCross of a purple-flowered and a white-flowered strain of peas; R stands for the gene for purple flowers and r for the gene for white flowers.

Mendel also crossbred varieties of peas that differed in two or more easily distinguishable traits. When a variety with yellow round seed was crossed to a green wrinkled-seed variety, the F_1 generation hybrids produced yellow round seed. Evidently, yellow (A) and round (B) are dominant traits, and green (a) and wrinkled (b) are recessive. By allowing the F_1 plants (genotype AaBb) to self-pollinate, Mendel obtained an F_2 generation of 315 yellow round, 101 yellow wrinkled, 108 green round, and 32 green wrinkled seeds, a ratio of approximately 9 : 3 : 3 : 1. The important point here is that the segregation of the colour (A–a) is independent of the segregation of the trait of seed surface (B–b). This is expected if the F_1 generation produces equal numbers of four kinds of gametes, carrying the four possible combinations of the parental genes: AB, Ab, aB, and ab. Random union of these gametes gives, then, the four phenotypes in a ratio 9 dominant–dominant : 3 recessive–dominant : 3 dominant–recessive : 1 recessive–recessive. Among these four phenotypic classes there must be nine different genotypes, a supposition that can be tested experimentally by raising a third hybrid generation. The predicted genotypes are actually found. Another test is by means of a backcross (or testcross); the F_1 hybrid (phenotype yellow round seed; genotype AaBb) is crossed to a double recessive plant (phenotype green wrinkled seed; genotype aabb). If the hybrid gives four kinds of gametes in equal numbers and if all the gametes of the double recessive are alike (ab), the predicted progeny of the backcross are yellow round, yellow wrinkled, green round, and green wrinkled seed in a ratio 1 : 1 : 1 : 1. This prediction is realized in experiments. When the varieties crossed differ in three genes, the F_1 hybrid forms 23, or eight, kinds of gametes (2n = kinds of gametes, n being the number of genes). The second generation of hybrids, the F_2, has 27 (33) genotypically distinct kinds of individuals but only eight different phenotypes. From these results and others, Mendel derived his second law: the law of recombination, or independent assortment of genes.

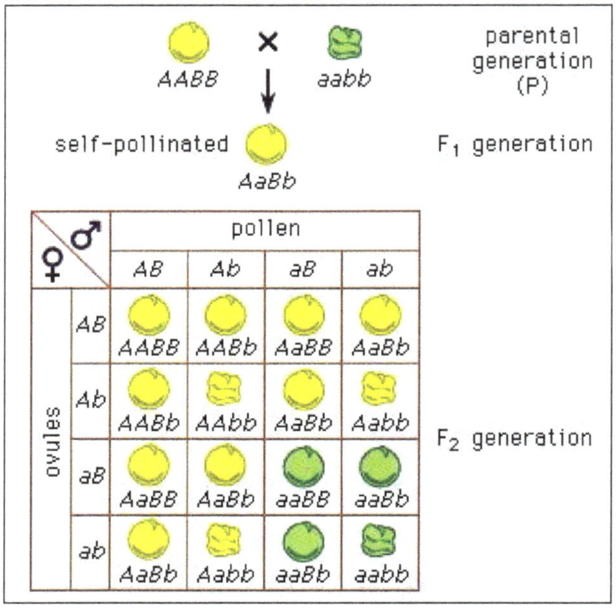

Mendel's law of independent assortmentCross of peas having yellow round seeds with peas having green wrinkled seeds. A stands for the gene for yellow and a for the gene for green; B stands for the gene for a round surface and b for the gene for a wrinkled surface.

Universality of Mendel's Laws

Although Mendel experimented with varieties of peas, his laws have been shown to apply to the inheritance of many kinds of characters in almost all organisms. In 1902 Mendelian inheritance was demonstrated in poultry and in mice. The following year, albinism became the first human trait shown to be a Mendelian recessive, with pigmented skin the corresponding dominant.

In 1902 and 1909, English physician Sir Archibald Garrod initiated the analysis of inborn errors of metabolism in humans in terms of biochemical genetics. Alkaptonuria, inherited as a recessive, is characterized by excretion in the urine of large amounts of the substance called alkapton, or homogentisic acid, which renders the urine black on exposure to air. In normal (i.e., nonalkaptonuric) persons the homogentisic acid is changed to acetoacetic acid, the reaction being facilitated by an enzyme, homogentisic acid oxidase. Garrod advanced the hypothesis that this enzyme is absent or inactive in homozygous carriers of the defective recessive alkaptonuria gene; hence, the homogentisic acid accumulates and is excreted in the urine. Mendelian inheritance of numerous traits in humans has been studied since then.

In analyzing Mendelian inheritance, it should be borne in mind that an organism is not an aggregate of independent traits, each determined by one gene. A "trait" is really an abstraction, a term of convenience in description. One gene may affect many traits (a condition termed pleiotropic). The white gene in Drosophila flies is pleiotropic; it affects the colour of the eyes and of the testicular envelope in the males, the fecundity and the shape of the spermatheca in the females, and the longevity of both sexes. In humans many diseases caused by a single defective gene will have a variety of symptoms, all pleiotropic manifestations of the gene.

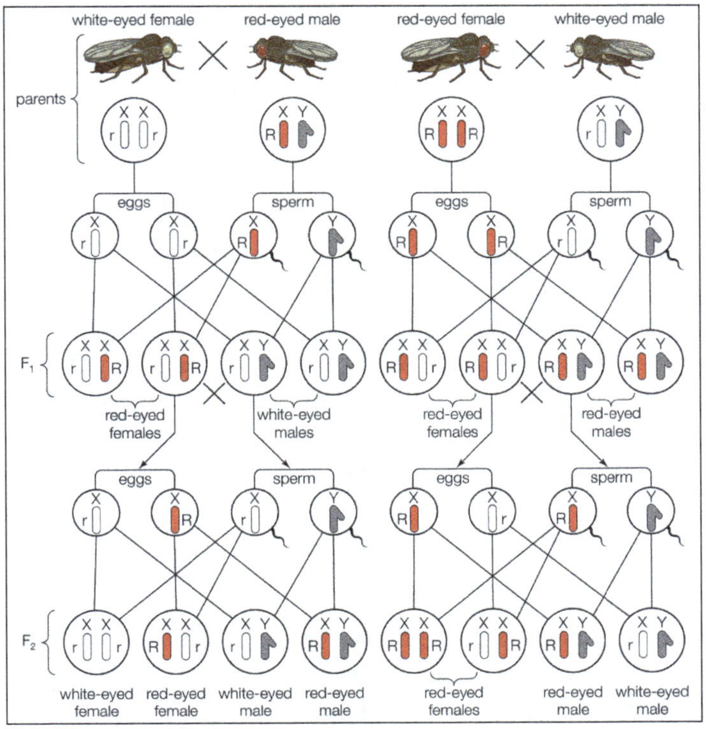

Sex-linked inheritance of white eyes in Drosophila flies.

Allelic Interactions

Dominance Relationships

The operation of Mendelian inheritance is frequently more complex than in the case of the traits recorded by Mendel. In the first place, clear-cut dominance and recessiveness are by no means always found. When red- and white-flowered varieties of four-o'clock plants or snapdragons are crossed, for example, the F_1 hybrids have flowers of intermediate pink or rose colour, a situation that seems more explicable by the blending notion of inheritance than by Mendelian concepts. That the inheritance of flower colour is indeed due to Mendelian mechanisms becomes apparent when the F_1 hybrids are allowed to cross, yielding an F_2 generation of red-, pink-, and white-flowered plants in a ratio of 1 red : 2 pink : 1 white. Obviously the hereditary information for the production of red and white flowers had not been blended away in the first hybrid generation, as flowers of these colours were produced in the second generation of hybrids.

The apparent blending in the F_1 generation is explained by the fact that the gene alleles that govern flower colour in four-o'clocks show an incomplete dominance relationship. Suppose then that a gene allele R1 is responsible for red flowers and R_2 for white; the homozygotes R_1R_1 and R_2R_2 are red and white respectively, and the heterozygotes R_1R_2 have pink flowers. A similar pattern of lack of dominance is found in Shorthorn cattle. In diverse organisms, dominance ranges from complete (a heterozygote indistinguishable from one of the homozygotes) to incomplete (heterozygotes exactly intermediate) to excessive or overdominance (a heterozygote more extreme than either homozygote).

Another form of dominance is one in which the heterozygote displays the phenotypic characteristics of both alleles. This is called codominance; an example is seen in the MN blood group system of human beings. MN blood type is governed by two alleles, M and N. Individuals who are homozygous for the M allele have a surface molecule (called the M antigen) on their red blood cells. Similarly, those homozygous for the N allele have the N antigen on the red blood cells. Heterozygotes—those with both alleles—carry both antigens.

Multiple Alleles

The traits discussed so far all have been governed by the interaction of two possible alleles. Many genes, however, are represented by multiple allelic forms within a population. (One individual, of course, can possess only two of these multiple alleles.) Human blood groups—in this case, the well-known ABO system—again provide an example. The gene that governs ABO blood types has three alleles: I^A, I^B, and I^O. I^A and I^B are codominant, but I^O is recessive. Because of the multiple alleles and their various dominance relationships, there are four phenotypic ABO blood types: type A (genotypes I^AI^A and I^AI^O), type B (genotypes I^BI^B and I^BI^O), type AB (genotype I^AI^B), and type O (genotype I^OI^O).

Gene Interactions

Many individual traits are affected by more than one gene. For example, the coat colour in many mammals is determined by numerous genes interacting to produce the result. The great variety of colour patterns in cats, dogs, and other domesticated animals is the result of different combinations of complexly interacting genes. The gradual unraveling of their modes of inheritance was one of the active fields of research in the early years of genetics.

Two or more genes may produce similar and cumulative effects on the same trait. In humans the skin-colour difference between so-called blacks and so-called whites is due to several (probably four or more) interacting pairs of genes, each of which increases or decreases the skin pigmentation by a relatively small amount.

Epistatic Genes

Some genes mask the expression of other genes just as a fully dominant allele masks the expression of its recessive counterpart. A gene that masks the phenotypic effect of another gene is called an epistatic gene; the gene it subordinates is the hypostatic gene. The gene for albinism (lack of pigment) in humans is an epistatic gene. It is not part of the interacting skin-colour genes described above; rather, its dominant allele is necessary for the development of any skin pigment, and its recessive homozygous state results in the albino condition regardless of how many other pigment genes may be present. Albinism thus occurs in some individuals among dark- or intermediate-pigmented peoples as well as among light-pigmented peoples.

The presence of epistatic genes explains much of the variability seen in the expression of such dominantly inherited human diseases as Marfan syndrome and neurofibromatosis. Because of the effects of an epistatic gene, some individuals who inherit a dominant, disease-causing gene show only partial symptoms of the disease; some in fact may show no expression of the disease-causing gene, a condition referred to as nonpenetrance. The individual in whom such a nonpenetrant mutant gene exists will be phenotypically normal but still capable of passing the deleterious gene on to offspring, who may exhibit the full-blown disease.

Examples of epistasis abound in nonhuman organisms. In mice, as in humans, the gene for albinism has two variants: the allele for nonalbino and the allele for albino. The latter allele is unable to synthesize the pigment melanin. Mice, however, have another pair of alleles involved in melanin placement. These are the agouti allele, which produces dark melanization of the hair except for a yellow band at the tip, and the black allele, which produces melanization of the whole hair. If melanin cannot be formed (the situation in the mouse homozygous for the albino gene), neither agouti nor black can be expressed. Hence, homozygosity for the albinism gene is epistatic to the agouti/black alleles and prevents their expression.

Complementation

The phenomenon of complementation is another form of interaction between nonallelic genes. For example, there are mutant genes that in the homozygous state produce profound deafness in humans. One would expect that the children of two persons with such hereditary deafness would be deaf. This is frequently not the case, because the parents' deafness is often caused by different genes. Since the mutant genes are not alleles, the child becomes heterozygous for the two genes and hears normally. In other words, the two mutant genes complement each other in the child. Complementation thus becomes a test for allelism. In the case of congenital deafness cited above, if all the children had been deaf, one could assume that the deafness in each of the parents was owing to mutant genes that were alleles. This would be more likely to occur if the parents were genetically related (consanguineous).

Polygenic Inheritance

The greatest difficulties of analysis and interpretation are presented by the inheritance of many quantitative or continuously varying traits. Inheritance of this kind produces variations in degree rather than in kind, in contrast to the inheritance of discontinuous traits resulting from single genes of major effect. The yield of milk in different breeds of cattle; the egg-laying capacity in poultry; and the stature, shape of the head, blood pressure, and intelligence in humans range in continuous series from one extreme to the other and are significantly dependent on environmental conditions. Crosses of two varieties differing in such characters usually give F1 hybrids intermediate between the parents. At first sight this situation suggests a blending inheritance through "blood" rather than Mendelian inheritance; in fact, it was probably observations of this kind of inheritance that suggested the folk idea of "blood theory."

It has, however, been shown that these characters are polygenic—i.e., determined by several or many genes, each taken separately producing only a slight effect on the phenotype, as small as or smaller than that caused by environmental influences on the same characters. That Mendelian segregation does take place with polygenes, as with the genes having major effects (sometimes called oligogenes), is shown by the variation among F2 and further-generation hybrids being usually much greater than that in the F1 generation. By selecting among the segregating progenies the desired variants—for example, individuals with the greatest yield, the best size, or a desirable behavior—it is possible to produce new breeds or varieties sometimes exceeding the parental forms. Hybridization and selection are consequently potent methods that have been used for improvement of agricultural plants and animals.

Polygenic inheritance also applies to many of the birth defects (congenital malformations) seen in humans. Although expression of the defect itself may be discontinuous (as in clubfoot, for example), susceptibility to the trait is continuously variable and follows the rules of polygenic inheritance. When a developmental threshold produced by a polygenically inherited susceptibility and a variety of environmental factors is exceeded, the birth defect results.

Heredity and Environment

Preformism and Epigenesist

A notion that was widespread among pioneer biologists in the 18th century was that the fetus, and hence the adult organism that develops from it, is preformed in the sex cells. Some early microscopists even imagined that they saw a tiny homunculus, a diminutive human figure, encased in the human spermatozoon. The development of the individual from the sex cells appeared deceptively simple: it was merely an increase in the size and growth of what was already present in the sex cells. The antithesis of the early preformation theories was theories of epigenesis, which claimed that the sex cells were structureless jelly and contained nothing at all in the way of rudiments of future organisms. The naive early versions of preformation and epigenesis had to be given up when embryologists showed that the embryo develops by a series of complex but orderly and gradual transformations. Darwin's "Provisional Hypothesis of Pangenesis" was distinctly preformistic; Weismann's theory of determinants in the germ plasm, as well as the early ideas about the relations between genes and traits, also tended toward preformism.

Heredity has been defined as a process that results in the progeny's resembling his parents. A further qualification of this definition states that what is inherited is a potential that expresses itself only after interacting with and being modified by environmental factors. In short, all phenotypic expressions have both hereditary and environmental components, the amount of each varying for different traits. Thus, a trait that is primarily hereditary (e.g., skin colour in humans) may be modified by environmental influences (e.g., suntanning). And conversely, a trait sensitive to environmental modifications (e.g., weight in humans) is also genetically conditioned. Organic development is preformistic insofar as a fertilized egg cell contains a genotype that conditions the events that may occur and is epigenetic insofar as a given genotype allows a variety of possible outcomes. These considerations should dispel the reluctance felt by many people to accept the fact that mental as well as physiological and physical traits in humans are genetically conditioned. Genetic conditioning does not mean that heredity is the "dice of destiny." At least in principle, but not invariably in practice, the development of a trait may be manipulated by changes in the environment.

Heritability

Although hereditary diseases and malformations are, unfortunately, by no means uncommon in the aggregate, no one of them occurs very frequently. The characteristics by which one person is distinguished from another—such as facial features, stature, shape of the head, skin, eye and hair colours, and voice—are not usually inherited in a clear-cut Mendelian manner, as are some hereditary malformations and diseases. This is not as strange as it may seem. The kinds of gene changes, or mutations, that produce morphological or physiological effects drastic enough to be clearly set apart from the more usual phenotypes are likely to cause diseases or malformations just because they are so drastic.

The variations that occur among healthy persons are, as a general rule, caused by polygenes with individually small effects. The same is true of individual differences among members of various animal and plant species. Even brown-blue eye colour in humans, which in many families behaves as if caused by two forms of a single gene (brown being dominant and blue recessive), is often blurred by minor gene modifiers of the pigmentation. Some apparently blue-eyed persons actually carry the gene for the brown eye colour, but several additional modifier genes decrease the amount of brown pigment in the iris. This type of genetic process can influence susceptibility to many diseases (e.g., diabetes) or birth defects (e.g., cleft lip—with or without cleft palate).

The question geneticists must often attempt to answer is how much of the observed diversity between persons or between individuals of any species is because of hereditary, or genotypic, variations and how much of it is because of environmental influences. Applied to human beings, this is sometimes referred to as the nature-nurture problem. With animals or plants the problem is evidently more easily soluble than it is with people. Two complementary approaches are possible. First, individual organisms or their progenies are raised in environments as uniform as can be provided, with food, temperature, light, humidity, etc., carefully controlled. The differences that persist between such individuals or progenies probably reflect genotypic differences. Second, individuals with similar or identical genotypes are placed in different environments. The phenotypic differences may then be ascribed to environmental induction. Experiments combining both approaches have been carried out on several species of plants that grow naturally at different altitudes, from sea level to the alpine zone of the Sierra Nevada in California. Young yarrow plants (Achillea) were cut into three parts, and the cuttings were replanted in experimental gardens at sea level, at mid-altitude (4,800 feet [1,460 metres]), and at high altitude (10,000 feet [3,050 metres]). It was observed that the plants native at sea level grow best in their native habitat, grow less well at mid-altitudes, and die at high altitudes. On the other hand, the alpine race survives and develops better at the high-altitude transplant station than it does at lower altitudes.

With organisms that cannot survive being cut into pieces and placed in controlled environments, a partitioning of the observed variability into genetic and environmental components may be attempted by other methods. Suppose that in a certain population individuals vary in stature, weight, or some other trait. These characters can be measured in many pairs of parents and in their progenies raised under different environmental conditions. If the variation is owing entirely to environment and not at all to heredity, then the expression of the character in the parents and in the offspring will show no correlation (heritability = zero). On the other hand, if the environment is unimportant and the character is uncomplicated by dominance, then the means of this character in the progenies will be the same as the means of the parents; with differences in the expression in females and in males taken into account, the heritability will equal unity. In reality, most heritabilities are found to lie between zero and one.

It is important to understand clearly the meaning of heritability estimates. They show that, given the range of the environments in which the experimental animals lived, one could predict the average body sizes in the progenies of pigs better than one could predict the average numbers of piglets in a litter. The heritability is, however, not an inherent or unchangeable property of each character. If one could make the environments more uniform, the heritabilities would rise, and with more-diversified environments they would decrease. Similarly, in populations that are more variable genetically, the heritabilities increase, and in genetically uniform ones, they decrease. In humans the situation is even more complex, because the environments of the parents and of their children are in many ways interdependent. Suppose, for example, that one wishes to study the heritability of stature, weight, or susceptibility to tuberculosis. The stature, weight, and liability to contract tuberculosis depend to some extent on the quality of nutrition and generally on the economic well-being of the family. If no allowance is made for this fact, the heritability estimates arrived at may be spurious; such heritabilities have indeed been claimed for such things as administrative, legal, or military talents and for social eminence in general. It is evident that having socially eminent parents makes it easier for the children to achieve such eminence also; biological heredity may have little or nothing to do with this.

A general conclusion from the evidence now available may be stated as follows: diversity in almost any trait—physical, physiological, or behavioral—owes in part to genetic variables and in part to environmental variables. In any array of environments, individuals with more nearly similar genetic endowments are likely to show a greater average resemblance than the carriers of more diverse genetic endowments. It is, however, also true that in different environments the carriers of similar genetic endowments may grow, develop, and behave in different ways.

Physical basis of Heredity

When Gregor Mendel formulated his laws of heredity, he postulated a particulate nature for the units of inheritance. What exactly these particles were he did not know. Today scientists understand not only the physical location of hereditary units (i.e., the genes) but their molecular composition as well. The unraveling of the physical basis of heredity makes up one of the most fascinating chapters in the history of biology.

Chromosomes and Genes

Each individual in a sexually reproducing species inherits two alleles for each gene, one from each parent. Furthermore, when such an individual forms sex cells, each of the resultant gametes receives one member of each allelic pair. The formation of gametes occurs through a process of cell division called meiosis. When gametes unite in fertilization, the double dose of hereditary material is restored, and a new individual is created. This individual, consisting at first of only one cell, grows via mitosis, a process of repeated cell divisions. Mitosis differs from meiosis in that each daughter cell receives a full copy of all the hereditary material found in the parent cell.

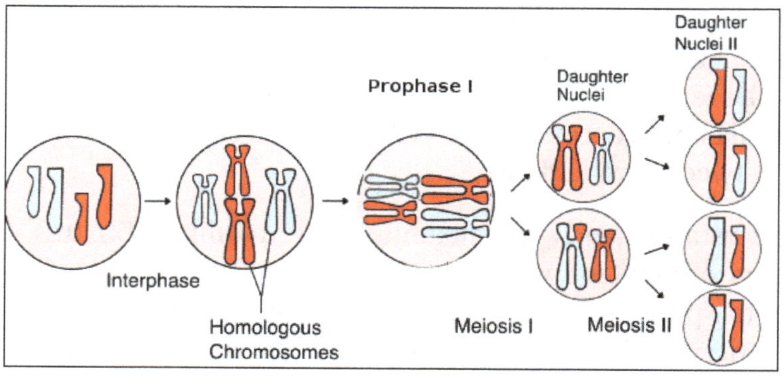

Cell Division: Mitosis and Meiosis.

It is apparent that the genes must physically reside in cellular structures that meet two criteria. First, these structures must be replicated and passed on to each generation of daughter cells during mitosis. Second, they must be organized into homologous pairs, one member of which is parceled out to each gamete formed during meiosis.

As early as 1848, biologists had observed that cell nuclei resolve themselves into small rodlike bodies during mitosis; later these structures were found to absorb certain dyes and so came to be called chromosomes (coloured bodies). During the early years of the 20th century, cellular studies using ordinary light microscopes clarified the behavior of chromosomes during mitosis and meiosis, which led to the conclusion that chromosomes are the carriers of genes.

Behavior of Chromosomes during Cell Division

During Mitosis

When the chromosomes condense during cell division, they have already undergone replication. Each chromosome thus consists of two identical replicas, called chromatids, joined at a point called the centromere. During mitosis the sister chromatids separate, one going to each daughter cell. Chromosomes thus meet the first criterion for being the repository of genes: they are replicated, and a full copy is passed to each daughter cell during mitosis.

During Meiosis

It was the behavior of chromosomes during meiosis, however, that provided the strongest evidence for their being the carriers of genes. In 1902 American scientist Walter S. Sutton reported on his observations of the action of chromosomes during sperm formation in grasshoppers. Sutton had observed that, during meiosis, each chromosome (consisting of two chromatids) becomes paired with a physically similar chromosome. These homologous chromosomes separate during meiosis, with one member of each pair going to a different cell. Assuming that one member of each homologous pair was of maternal origin and the other was paternally derived, here was an event that fulfilled the behavior of genes postulated in Mendel's first law.

It is now known that the number of chromosomes within the nucleus is usually constant in all individuals of a given species—for example, 46 in the human, 40 in the house mouse, 8 in the vinegar fly (Drosophila melanogaster; sometimes called fruit fly), 20 in corn (maize), 24 in the tomato, and 48 in the potato. In sexually reproducing organisms, this number is called the diploid number of chromosomes, as it represents the double dose of chromosomes received from two parents. The nucleus of a gamete, however, contains half this number of chromosomes, or the haploid number. Thus, a human gamete contains 23 chromosomes, while a Drosophila gamete contains four. Meiosis produces the haploid gametes.

The essential features of meiosis are shown in the diagram. For the sake of simplicity, the diploid parent cell is shown to contain a single pair of homologous chromosomes, one member of which is represented in blue (from the father) and the other in red (from the mother). At the leptotene stage the chromosomes appear as long, thin threads. At pachytene they pair, the corresponding portions of the two chromosomes lying side by side. The chromosomes then duplicate and contract into paired chromatids. At this stage the pair of chromosomes is known as a tetrad, as it consists of four chromatids. Also at this stage an extremely important event occurs: portions of the maternal and paternal chromosomes are exchanged. This exchange process, called crossing over, results in chromatids that include both paternal and maternal genes and consequently introduces new genetic combinations. The first meiotic division separates the chromosomal tetrads, with the paternal chromosome (whose chromatids now contain some maternal genes) going to one cell and the maternal chromosome (containing some paternal genes) going to another cell. During the second meiotic division the chromatids separate. The original diploid cell has thus given rise to four haploid gametes (only two of which are shown in the diagram). Not only has a reduction in chromosome number occurred, but the resulting single member of each homologous chromosome pair may be a new combination (through crossing over) of genes present in the original diploid cell.

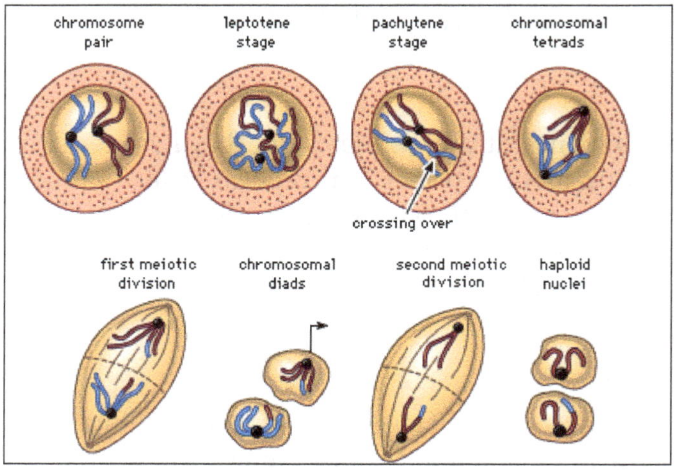

Behavior of chromosomes at meiosis.

Suppose that the red chromosome shown in the diagram carries the gene for albinism, and the blue chromosome carries the gene for dark pigmentation. It is evident that the two gene alleles will undergo segregation at meiosis and that one-half of the gametes formed will contain the albino gene and the other half the pigmentation gene. Following the scheme in the diagram, random combination of the gametes with the albino gene and the pigmentation gene will give two kinds of homozygotes and one kind of heterozygote in a ratio of 1 : 1 : 2. Mendel's law of segregation is thus the outcome of chromosome behavior at meiosis. The same is true of the second law, that of independent assortment.

Consider the inheritance of two pairs of genes, such as Mendel's factors for seed coloration and seed surface in peas; these genes are located on different pairs of chromosomes. Since maternal and paternal members of different chromosome pairs are assorted independently, so are the genes they contain. This explains, in part, the genetic variety seen among the progeny of the same pair of parents. As stated above, humans have 46 chromosomes in the body cells and in the cells (oogonia and spermatogonia) from which the sex cells arise. At meiosis these 46 chromosomes form 23 pairs, one of the chromosomes of each pair being of maternal and the other of paternal origin. Independent assortment is, then, capable of producing 2^{23}, or 8,388,608, kinds of sex cells with different combinations of the grandmaternal and grandpaternal chromosomes. Since each parent has the potentiality of producing 2^{23} kinds of sex cells, the total number of possible combinations of the grandparental chromosomes is $2^{23} \times 2^{23} = 2^{46}$. The population of the world is now more than 6 billion persons, or approximately 2^{32} persons. It is therefore certain that only a tiny fraction of the potentially possible chromosome and gene combinations can ever be realized. Yet even 246 is an underestimate of the variety potentially possible. The grandmaternal and grandpaternal members of the chromosome pairs are not indivisible units. Each chromosome carries many genes, and the chromosome pairs exchange segments at meiosis through the process of crossing over. This is evidence that the genes rather than the chromosomes are the units of Mendelian segregation.

Linkage of Traits

Simple Linkage

As pointed out above, the random assortment of the maternal and paternal chromosomes at meiosis is the physical basis of the independent assortment of genes and of the traits they control. This is the basis of the second law of Mendel. The number of the genes in a sex cell is, however, much greater than that of the chromosomes. When two or more genes are borne on the same chromosome, these genes may not be assorted independently; such genes are said to be linked. When a Drosophila fly homozygous for a normal gray body and long wings is crossed with one having a black body and vestigial wings, the F_1 consists of hybrid gray, long-winged flies. Gray body (B) is evidently dominant over black body (b), and long wing (V) is dominant over vestigial wing (v). Now consider a backcross of the heterozygous F_1 males to double-recessive black-vestigial females (bbvv). Independent assortment would be expected to give in the progeny of the backcross the following: 1 gray-long : 1 gray-vestigial : 1 black-long : 1 black-vestigial. In reality, only gray-long and black-vestigial flies are produced, in approximately equal numbers; the genes remain linked in the same combinations in which they were found in the parents. The backcross of the heterozygous F_1 females to double-recessive males gives a somewhat different result: 42 percent each of gray-long and black-vestigial flies and about 8 percent each of black-long and gray-vestigial classes. In sum, 84 percent of the progeny have the parental combinations of traits, and 16 percent have the traits recombined. The interpretation of these results given in 1911 by the American geneticist Thomas Hunt Morgan laid the foundation of the theory of linear arrangement of genes in the chromosomes.

Traits that exhibit linkage in experimental crosses (such as black body and vestigial wings) are determined by genes located in the same chromosome. As more and more genes became known in Drosophila, they fell neatly into four linkage groups corresponding to the four pairs of the chromosomes this species possesses. One linkage group consists of sex-linked genes, located in the X chromosome; of the three remaining linkage groups, two have many more genes than the remaining one, which corresponds to the presence of two pairs of large chromosomes and one pair of tiny dot-like chromosomes. The numbers of linkage groups in other organisms are equal to or smaller than the numbers of the chromosomes in the sex cells—e.g., 10 linkage groups and 10 chromosomes in corn, 19 linkage groups and 20 chromosomes in the house mouse, and 23 linkage groups and 23 chromosomes in the human.

The linkage of the genes black and vestigial in Drosophila is complete in heterozygous males, while in the progeny of females there appear about 17 percent of recombination classes. With very rare exceptions, the linkage of all genes belonging to the same linkage group is complete in Drosophila males, while in the females different pairs of genes exhibit all degrees of linkage from complete (no recombination) to 50 percent (random assortment). Morgan's inference was that the degree of linkage depends on physical distance between the genes in the chromosome: the closer the genes, the tighter the linkage and vice versa. Furthermore, Morgan perceived that the chiasmata (crosses that occur in meiotic chromosomes) indicate the mechanism underlying the phenomena of linkage and crossing over. The chromosomes at meiosis, the maternal and paternal chromosomes (represented in blue and red) cross over and exchange segments, so a chromosome emerging from the process of meiosis may consist of some maternal (grandmaternal) and some paternal (grandpaternal) sections. If the probability of crossing over taking place is uniform along the length of a chromosome (which was later shown to be not quite true), then genes close together will be recombined less frequently than those far apart.

This realization opened an opportunity to map the arrangement of the genes and the estimated distances between them in the chromosome by studying the frequencies of recombination of various traits in the progenies of hybrids. In other words, the linkage maps of the chromosomes are really summaries of many statistical observations on the outcomes of hybridization experiments. In principle at least, such maps could be prepared even if the chromosomes, not to speak of the chiasmata at meiosis, were unknown. But an interesting and relevant fact is that in Drosophila males the linkage of the genes in the same chromosome is complete, and observations under the microscope show that no chiasmata are formed in the chromosomes at meiosis. In most organisms, including humans, chiasmata are seen in the meiotic chromosomes in both sexes, and observations on hybrid progenies show that recombination of linked genes occurs also in both sexes.

Chromosome maps exist for the Drosophila fly, corn, the house mouse, the bread mold Neurospora crassa, and some bacteria and bacteriophages (viruses that infect bacteria). Until quite late in the 20th century, the mapping of human chromosomes presented a particularly difficult problem: experimental crosses could not be arranged in humans, and only a few linkages could be determined by analysis of unique family histories. However, the development of recombinant DNA technology provided new understanding of human genetic processes and new methods of research. Using the techniques of recombinant DNA technology, hundreds of genes have been mapped to the human chromosomes and many linkages established.

Sex Linkage

The male of many animals has one chromosome pair, the sex chromosomes, consisting of unequal members called X and Y. At meiosis the X and Y chromosomes first pair then disjoin and pass to different cells. One-half of the gametes (spermatozoa) formed contain the X chromosome and the other half the Y. The female has two X chromosomes; all egg cells normally carry a single X. The eggs fertilized by X-bearing spermatozoa give females (XX), and those fertilized by Y-bearing spermatozoa give males (XY).

The genes located in the X chromosomes exhibit what is known as sex-linkage or crisscross inheritance. This is because of a crucial difference between the paired sex chromosomes and the other pairs of chromosomes (called autosomes). The members of the autosome pairs are truly homologous; that is, each member of a pair contains a full complement of the same genes (albeit, perhaps, in different allelic forms). The sex chromosomes, on the other hand, do not constitute a homologous pair, as the X chromosome is much larger and carries far more genes than does the Y. Consequently, many recessive alleles carried on the X chromosome of a male will be expressed just as if they were dominant, for the Y chromosome carries no genes to counteract them. The classic case of sex-linked inheritance, described by Morgan in 1910, is that of the white eyes in Drosophila. White-eyed females crossed to males with the normal red eye colour produce red-eyed daughters and white-eyed sons in the F_1 generation and equal numbers of white-eyed and red-eyed females and males in the F_2 generation. The cross of red-eyed females to white-eyed males gives a different result: both sexes are red-eyed in F_1 and the females in the F_2 generation are red-eyed, half the males are red-eyed, and the other half white-eyed. As interpreted by Morgan, the gene that determines the red or white eyes is borne on the X chromosome, and the allele for red eye is dominant over that for white eye. Since a male receives its single X chromosome from his mother, all sons of white-eyed females also have white eyes. A female inherits one X chromosome from her mother and the other X from her father. Red-eyed females may have genes for red eyes in both of their X chromosomes (homozygotes), or they may have one X with the gene for red and the other for white (heterozygotes). In the progeny of heterozygous females, one-half of the sons will receive the X chromosome with the gene for white and will have white eyes, and the other half will receive the X with the gene for red eyes. The daughters of the heterozygous females crossed with white-eyed males will have either two X chromosomes with the gene for white—and hence have white eyes—or one X with the gene for white and the other X with the gene for red and will be red-eyed heterozygotes.

In humans, red-green colour blindness and hemophilia are among many traits showing sex-linked inheritance and are consequently due to genes borne in the X chromosome.

In some animals—birds, butterflies and moths, some fish, and at least some amphibians and reptiles—the chromosomal mechanism of sex determination is a mirror image. The male has two X chromosomes and the female an X and Y chromosome. Here the spermatozoa all have an X chromosome; the eggs are of two kinds, some with X and others with Y chromosomes, usually in equal numbers. The sex of the offspring is then determined by the egg rather than by the spermatozoon. Sex-linked inheritance is altered correspondingly. A male homozygous for a sex-linked recessive trait crossed to a female with the dominant one gives, in the F_1 generation, daughters with the recessive trait and heterozygous sons with the corresponding dominant trait. The F_2 generation has recessive and dominant females and males in equal numbers. A male with a dominant trait crossed to a female with a recessive trait gives uniformly dominant F_1 and a segregation in a ratio of 2 dominant males : 1 dominant female : 1 recessive female.

Observations on pedigrees or experimental crosses show that certain traits exhibit sex-linked inheritance; the behavior of the X chromosomes at meiosis is such that the genes they carry may be expected to exhibit sex-linkage. This evidence still failed to convince some skeptics that the genes for the sex-linked traits were in fact borne in certain chromosomes seen under the microscope. An experimental proof was furnished in 1916 by American geneticist Calvin Blackman Bridges. As stated above, white-eyed Drosophila females crossed to red-eyed males usually produce red-eyed female and white-eyed male progeny. Among thousands of such "regular" offspring, there are occasionally found exceptional white-eyed females and red-eyed males. Bridges constructed the following working hypothesis. Suppose that, during meiosis in the female, gametogenesis occasionally goes wrong, and the two X chromosomes fail to disjoin. Exceptional eggs will then be produced, carrying two X chromosomes and eggs carrying none. An egg with two X chromosomes coming from a white-eyed female fertilized by a spermatozoon with a Y chromosome will give an exceptional white-eyed female. An egg with no X chromosome fertilized by a spermatozoon with an X chromosome derived from a red-eyed father will yield an exceptional red-eyed male. This hypothesis can be rigorously tested. The exceptional white-eyed females should have not only the two X chromosomes but also a Y chromosome, which normal females do not have. The exceptional males should, on the other hand, lack a Y chromosome, which normal males do have. Both predictions were verified by examination under a microscope of the chromosomes of exceptional females and males. The hypothesis also predicts that exceptional eggs with two X chromosomes fertilized by X-bearing spermatozoa must give individuals with three X chromosomes; such individuals were later identified by Bridges as poorly viable "superfemales." Exceptional eggs with no Xs, fertilized by Y-bearing spermatozoa, will give zygotes without X chromosomes; such zygotes die in early stages of development.

Chromosomal Aberrations

The chromosome set of a species remains relatively stable over long periods of time. However, within populations there can be found abnormalities involving the structure or number of chromosomes. These alterations arise spontaneously from errors in the normal processes of the cell. Their consequences are usually deleterious, giving rise to individuals who are unhealthy or sterile, though in rare cases alterations provide new adaptive opportunities that allow evolutionary change to occur. In fact, the discovery of visible chromosomal differences between species has given rise to the belief that radical restructuring of chromosome architecture has been an important force in evolution.

Changes in Chromosome Structure

Two important principles dictate the properties of a large proportion of structural chromosomal changes. The first principle is that any deviation from the normal ratio of genetic material in the genome results in genetic imbalance and abnormal function. In the normal nuclei of both diploid and haploid cells, the ratio of the individual chromosomes to one another is 1:1. Any deviation from this ratio by addition or subtraction of either whole chromosomes or parts of chromosomes results in genomic imbalance. The second principle is that homologous chromosomes go to great lengths to pair at meiosis. The tightly paired homologous regions are joined by a ladderlike longitudinal structure called the synaptonemal complex. Homologous regions seem to be able to find each other and form a synaptonemal complex whether or not they are part of normal chromosomes. Therefore, when structural changes occur, not only are the resulting pairing formations highly characteristic of that type of structural change but they also dictate the packaging of normal and abnormal chromosomes into the gametes and subsequently into the progeny.

Deletions

The simplest, but perhaps most damaging, structural change is a deletion—the complete loss of a part of one chromosome. In a haploid cell this is lethal, because part of the essential genome is lost. However, even in diploid cells deletions are generally lethal or have other serious consequences. In a diploid a heterozygous deletion results in a cell that has one normal chromosome set and another set that contains a truncated chromosome. Such cells show genomic imbalance, which increases in severity with the size of the deletion. Another potential source of damage is that any recessive, deleterious, or lethal alleles that are in the normal counterpart of the deleted region will be expressed in the phenotype. In humans, cri-du-chat syndrome is caused by a heterozygous deletion at the tip of the short arm of chromosome 5. Infants are born with this condition as the result of a deletion arising in parental germinal tissues or even in sex cells. The manifestations of this deletion, in addition to the "cat cry" that gives the syndrome its name, include severe intellectual disability and an abnormally small head.

Duplications

A heterozygous duplication (an extra copy of some chromosome region) also results in a genomic imbalance with deleterious consequences. Small duplications within a gene can arise spontaneously. Larger duplications can be caused by crossovers following asymmetrical chromosome pairing or by meiotic irregularities resulting from other types of altered chromosome structures. If a duplication becomes homozygous, it can provide the organism with an opportunity to acquire new genetic functions through mutations within the duplicate copy.

Inversions

An inversion occurs when a chromosome breaks in two places and the region between the break rotates 180° before rejoining with the two end fragments. If the inverted segment contains the centromere (i.e., the point where the two chromatids are joined), the inversion is said to be pericentric; if not, it is called paracentric. Inversions do not result in a gain or loss of genetic material, and they have deleterious effects only if one of the chromosomal breaks occurs within an essential gene or if the function of a gene is altered by its relocation to a new chromosomal neighbourhood (called the position effect). However, individuals who are heterozygous for inversions produce aberrant meiotic products along with normal products. The only way uninverted and inverted segments can pair is by forming an inversion loop. If no crossovers occur in the loop, half of the gametes will be normal and the other half will contain an inverted chromosome. If a crossover does occur within the loop of a paracentric inversion, a chromosome bridge and an acentric chromosome (i.e., a chromosome without a centromere) will be formed, and this will give rise to abnormal meiotic products carrying deletions, which are inviable. In a pericentric inversion, a crossover within the loop does not result in a bridge or an acentric chromosome, but inviable products are produced carrying a duplication and a deletion.

Translocations

If a chromosome break occurs in each of two nonhomologous chromosomes and the two breaks rejoin in a new arrangement, the new segment is called a translocation. A cell bearing a heterozygous translocation has a full set of genes and will be viable unless one of the breaks causes damage within a gene or if there is a position effect on gene function. However, once again the pairing properties of the chromosomes at meiosis result in aberrant meiotic products. Specifically, half of the products are deleted for one of the chromosome regions that changed positions and half of the products are duplicated for the other. These duplications and deletions usually result in inviability, so translocation heterozygotes are generally semisterile ("half-sterile").

Changes in Chromosome Number

Two types of changes in chromosome numbers can be distinguished: a change in the number of whole chromosome sets (polyploidy) and a change in chromosomes within a set (aneuploidy).

Polyploids

An individual with additional chromosome sets is called a polyploid. Individuals with three sets of chromosomes (triploids, 3n) or four sets of chromosomes (tetraploids, 4n) are polyploid derivatives of the basic diploid (2n) constitution. Polyploids with odd numbers of sets (e.g., triploids) are sterile, because homologous chromosomes pair only two by two, and the extra chromosome moves randomly to a cell pole, resulting in highly unbalanced, nonfunctional meiotic products. It is for this reason that triploid watermelons are seedless. However, polyploids with even numbers of chromosome sets can be fertile if orderly two-by-two chromosome pairing occurs.

Though two organisms from closely related species frequently hybridize, the chromosomes of the fusing partners are different enough that the two sets do not pair at meiosis, resulting in sterile offspring. However, if by chance the number of chromosome sets in the hybrid accidentally duplicates, a pairing partner for each chromosome will be produced, and the hybrid will be fertile. These chromosomally doubled hybrids are called allotetraploids. Bread wheat, which is hexaploid (6n) due to several natural spontaneous hybridizations, is an example of an allotetraploid. Some polyploid plants are able to produce seeds through an asexual type of reproduction called apomixis; in such cases, all progeny are identical to the parent. Polyploidy does arise spontaneously in humans, but all polyploids either abort in utero or die shortly after birth.

Aneuploids

Some cells have an abnormal number of chromosomes that is not a whole multiple of the haploid number. This condition is called aneuploidy. Most aneuploids arise by nondisjunction, a failure of homologous chromosomes to separate at meiosis. When a gamete of this type is fertilized by a normal gamete, the zygotes formed will have an unequal distribution of chromosomes. Such genomic imbalance results in severe abnormalities or death. Only aneuploids involving small chromosomes tend to survive and even then only with an aberrant phenotype.

The most common form of aneuploidy in humans results in Down syndrome, a suite of specific disorders in individuals possessing an extra chromosome 21 (trisomy 21). The symptoms of Down syndrome include intellectual disability, severe disorders of internal organs such as the heart and kidneys, up-slanted eyes, an enlarged tongue, and abnormal dermal ridge patterns on the fingers, palms, and soles. Other forms of aneuploidy in humans result from abnormal numbers of sex chromosomes. Turner syndrome is a condition in which females have only one X chromosome. Symptoms may include short stature, webbed neck, kidney or heart malformations, underdeveloped sex characteristics, or sterility. Klinefelter syndrome is a condition in which males have one extra female sex chromosome, resulting in an XXY pattern. (Other, less frequent, chromosomal patterns include XXXY, XXXXY, XXYY, and XXXYY.) Symptoms of Klinefelter syndrome may include sterility, a tall physique, lack of secondary sex characteristics, breast development, and learning disabilities.

The data accumulated by scientists of the early 20th century provided compelling evidence that chromosomes are the carriers of genes. But the nature of the genes themselves remained a mystery, as did the mechanism by which they exert their influence. Molecular genetics—the study of the structure and function of genes at the molecular level—provided answers to these fundamental questions.

DNA as the Agent of Heredity

In 1869 Swiss chemist Johann Friedrich Miescher extracted a substance containing nitrogen and phosphorus from cell nuclei. The substance was originally called nuclein, but it is now known as deoxyribonucleic acid, or DNA. DNA is the chemical component of the chromosomes that is chiefly responsible for their staining properties in microscopic preparations. Since the chromosomes of eukaryotes contain a variety of proteins in addition to DNA, the question naturally arose whether the nucleic acids or the proteins, or both together, were the carriers of the genetic information. Until the early 1950s most biologists were inclined to believe that the proteins were the chief carriers of heredity. Nucleic acids contain only four different unitary building blocks, but proteins are made up of 20 different amino acids. Proteins therefore appeared to have a greater diversity of structure, and the diversity of the genes seemed at first likely to rest on the diversity of the proteins.

Evidence that DNA acts as the carrier of the genetic information was first firmly demonstrated by exquisitely simple microbiological studies. In 1928 English bacteriologist Frederick Griffith was studying two strains of the bacterium Streptococcus pneumoniae; one strain was lethal to mice (virulent) and the other was harmless (avirulent). Griffith found that mice inoculated with either the heat-killed virulent bacteria or the living avirulent bacteria remained free of infection, but mice inoculated with a mixture of both became infected and died. It seemed as if some chemical "transforming principle" had transferred from the dead virulent cells into the avirulent cells and changed them. In 1944 American bacteriologist Oswald T. Avery and his coworkers found that the transforming factor was DNA. Avery and his research team obtained mixtures from heat-killed virulent bacteria and inactivated either the proteins, polysaccharides (sugar subunits), lipids, DNA, or RNA (ribonucleic acid, a close chemical relative of DNA) and added each type of preparation individually to avirulent cells. The only molecular class whose inactivation prevented transformation to virulence was DNA. Therefore, it seemed that DNA, because it could transform, must be the hereditary material.

A similar conclusion was reached from the study of bacteriophages, viruses that attack and kill bacterial cells. From a host cell infected by one bacteriophage, hundreds of bacteriophage progeny are produced. In 1952 American biologists Alfred D. Hershey and Martha Chase prepared two populations of bacteriophage particles. In one population, the outer protein coat of the bacteriophage was labeled with a radioactive isotope; in the other, the DNA was labeled. After allowing both populations to attack bacteria, Hershey and Chase found that only when DNA was labeled did the progeny bacteriophage contain radioactivity. Therefore, they concluded that DNA is injected into the bacterial cell, where it directs the synthesis of numerous complete bacteriophages at the expense of the host. In other words, in bacteriophages DNA is the hereditary material responsible for the fundamental characteristics of the virus.

Today the genetic makeup of most organisms can be transformed using externally applied DNA, in a manner similar to that used by Avery for bacteria. Transforming DNA is able to pass through cellular and nuclear membranes and then integrate into the chromosomal DNA of the recipient cell. Furthermore, using modern DNA technology, it is possible to isolate the section of chromosomal DNA that constitutes an individual gene, manipulate its structure, and reintroduce it into a cell to cause changes that show beyond doubt that the DNA is responsible for a large part of the overall characteristics of an organism. For reasons such as these, it is now accepted that, in all living organisms, with the exception of some viruses, genes are composed of DNA.

Structure and Composition of DNA

The remarkable properties of the nucleic acids, which qualify these substances to serve as the carriers of genetic information, have claimed the attention of many investigators. The groundwork was laid by pioneer biochemists who found that nucleic acids are long chainlike molecules, the backbones of which consist of repeated sequences of phosphate and sugar linkages—ribose sugar in RNA and deoxyribose sugar in DNA. Attached to the sugar links in the backbone are two kinds of nitrogenous bases: purines and pyrimidines. The purines are adenine (A) and guanine (G) in both DNA and RNA; the pyrimidines are cytosine (C) and thymine (T) in DNA and cytosine (C) and uracil (U) in RNA. A single purine or pyrimidine is attached to each sugar, and the entire phosphate-sugar-base subunit is called a nucleotide. The nucleic acids extracted from different species of animals and plants have different proportions of the four nucleotides. Some are relatively richer in adenine and thymine, while others have more guanine and cytosine. However, it was found by biochemist Erwin Chargaff that the amount of A is always equal to T, and the amount of G is always equal to C.

With the general acceptance of DNA as the chemical basis of heredity in the early 1950s, many scientists turned their attention to determining how the nitrogenous bases fit together to make up a threadlike molecule. The structure of DNA was determined by American geneticist James Watson and British biophysicist Francis Crick in 1953. Watson and Crick based their model largely on the research of British physicists Rosalind Franklin and Maurice Wilkins, who analyzed X-ray diffraction patterns to show that DNA is a double helix. The findings of Chargaff suggested to Watson and Crick that adenine was somehow paired with thymine and that guanine was paired with cytosine.

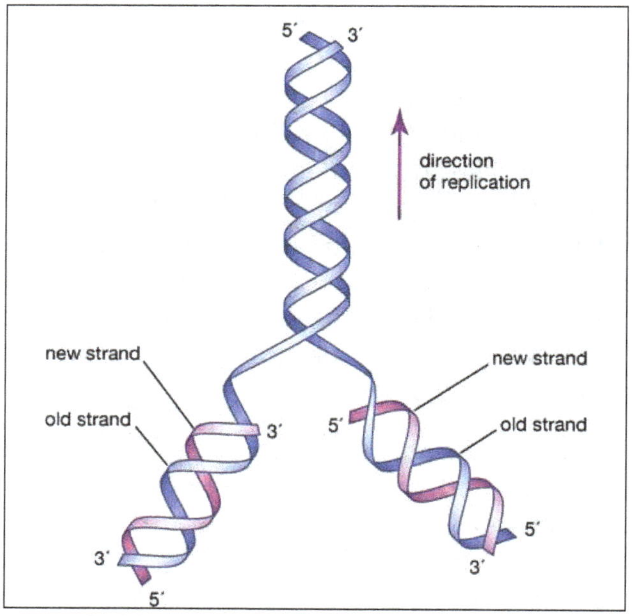

The initial proposal of the structure of DNA by James Watson and Francis Crick, which was accompanied by a suggestion on the means of replication.

Using this information, Watson and Crick came up with their now-famous model showing DNA as a double helix composed of two intertwined chains of nucleotides, in which the adenines of one chain are linked to the thymines of the other, and the guanines in one chain are linked to the cytosines of the other. The structure resembles a ladder that has been twisted into a spiral shape: the sides of the ladder are composed of sugar and phosphate groups, and the rungs are made up of the paired nitrogenous bases. By making a wire model of the structure, it became clear that the only way the model could conform to the requirements of the molecular dimensions of DNA was if A always paired with T and G with C; in fact, the A-T and G-C pairs showed a satisfying lock-and-key fit. Although most of the bonds in DNA are strong covalent bonds, the A-T and G-C bonds are weak hydrogen bonds. However, multiple hydrogen bonds along the centre of the molecule confer enough stability to hold the two strands together.

The two strands of Watson and Crick's double helix were antiparallel; that is, the nucleotides were arranged in opposite orientation. This can be visualized if the L shape of a nucleotide is imagined to be a sock: the neck of the sock is the nitrogenous base, the toe is the phosphate group, and the heel is the sugar group. The nucleotide chain would then be a string of socks attached heel to toe, with the necks pointing inward toward the centre of the DNA molecule. In one strand the arrangement of the sugar-phosphate backbone would be toe-heel-toe-heel and so on, and in the other strand in the same direction the arrangement would be heel-toe-heel-toe. Chemically, the heel is the 3′-hydroxyl end and the toe is the 5′-phosphate end. (These names are derived from the carbon atoms through which the sugar-phosphate linkage is made.) Therefore, one DNA strand runs from $5' \rightarrow 3'$ (five prime to three prime), whereas the other runs from $3' \rightarrow 5'$.

Watson and Crick noted that their proposed DNA structure fulfilled two necessary features of a hereditary molecule. First, a hereditary molecule must be capable of replication so that the information can be passed on to the next generation; therefore, Watson and Crick hypothesized that, if the two halves of the double helix could separate, they could act as templates for the synthesis of two identical double helices. Second, a hereditary molecule must contain information to guide the development of a complete organism; therefore, Watson and Crick speculated that the sequence of nucleotides might represent coded information of this sort. Subsequent research showed that their speculations on both points were correct.

DNA Replication

The Watson-Crick model of the structure of DNA suggested at least three different ways that DNA might self-replicate. The experiments of Matthew Meselson and Franklin Stahl on the bacterium Escherichia coli in 1958 suggested that DNA replicates semiconservatively. Meselson and Stahl grew bacterial cells in the presence of ^{15}N, a heavy isotope of nitrogen, so that the DNA of the cells contained ^{15}N. These cells were then transferred to a medium containing the normal isotope of nitrogen, ^{14}N, and allowed to go through cell division. The researchers were able to demonstrate that, in the DNA molecules of the daughter cells, one strand contained only ^{15}N, and the other strand contained ^{14}N. This is precisely what is expected by the semiconservative mode of replication, in which the original DNA molecules should separate into two template strands containing ^{15}N, and the newly aligned nucleotides should all contain ^{14}N.

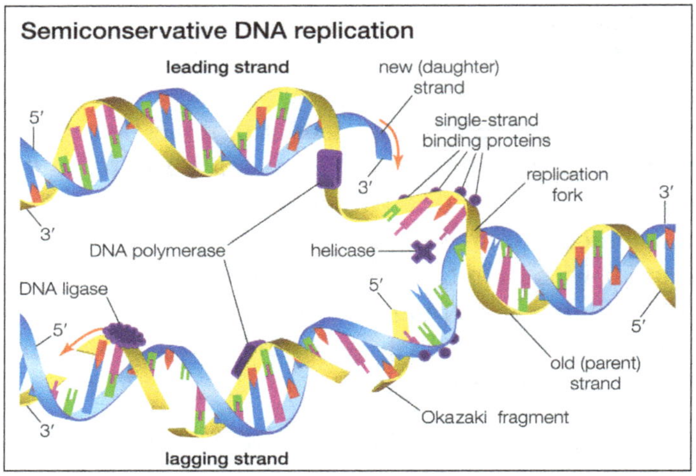

In semiconservative DNA replication an existing DNA molecule is separated into two template strands. New nucleotides align with and bind to the nucleotides of the existing strands, thus forming two DNA molecules that are identical to the original DNA molecule.

The hooking together of free nucleotides in the newly synthesized strand takes place one nucleotide at a time in the 5′ → 3′ direction. An incoming free nucleotide pairs with the complementary nucleotide on the template strand, and then the 5′ end of the free nucleotide is covalently joined to the 3′ end of a nucleotide already in place. The process is then repeated. The result is a nucleotide chain, referred to chemically as a nucleotide polymer or a polynucleotide. Of course the polymer is not a random polymer; its nucleotide sequence has been directed by the nucleotide sequence of the template strand. It is this templating process that enables hereditary information to be replicated accurately and passed down through the generations. In a very real way, human DNA has been replicated in a direct line of descent from the first vertebrates that evolved hundreds of millions of years ago.

DNA replication starts at a site on the DNA called the origin of replication. In higher organisms, replication begins at multiple origins of replication and moves along the DNA in both directions outward from each origin, creating two replication "forks." The events at both replication forks are identical. In order for DNA to replicate, however, the two strands of the double helix first must be unwound from each other. A class of enzymes called DNA topoisomerases removes helical twists by cutting a DNA strand and then resealing the cut. Enzymes called helicases then separate the two strands of the double helix, exposing two template surfaces for the alignment of free nucleotides. Beginning at the origin of replication, a complex enzyme called DNA polymerase moves along the DNA molecule, pairing nucleotides on each template strand with free complementary nucleotides. Because of the antiparallel nature of the DNA strands, new strand synthesis is different on each template. On the 3′ → 5′ template strand, polymerization proceeds in the 5′ → 3′ direction, and this growing strand is called the leading strand. However, polymerization must be carried out differently on the 5′ → 3′ template strand because nucleotides cannot be assembled in the 3′ → 5′ direction. Here short sequences of RNA are polymerized on the template. These sequences act as primers to which the DNA polymerase can add nucleotides in the 5′ → 3′ direction but in the opposite direction in which synthesis is proceeding on the lagging strand. The DNA polymerase hence makes short segments of DNA called Okazaki fragments in the "wrong" direction. For this reason the strand synthesized on the 5′ → 3′ template strand is called the lagging strand. Later, the RNA primers are removed and the Okazaki fragments are joined. This RNA priming system cannot be used to synthesize the very end of the 3′ → 5′ strand; once the last RNA primer is removed, synthesis cannot continue over the remaining gap. To overcome this obstacle, the enzyme telomerase adds multiple copies of a nucleotide sequence to the end of the DNA strand to allow completion of replication. Despite the peculiar events on the lagging strand, the entire DNA strand is eventually polymerized, and the two daughter DNA molecules thus produced are identical.

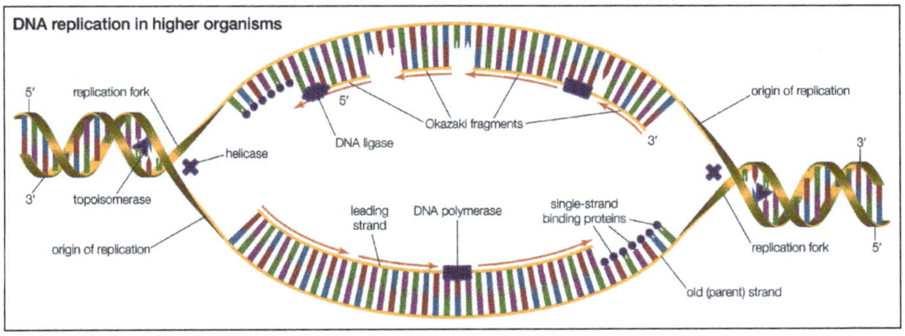

DNA replication in higher organisms begins at multiple origins of replication and progresses in two directions.

Expression of the Genetic Code: Transcription and Translation

DNA represents a type of information that is vital to the shape and form of an organism. It contains instructions in a coded sequence of nucleotides, and this sequence interacts with the environment to produce form—the living organism with all of its complex structures and functions. The form of an organism is largely determined by protein. A large proportion of what we see when we observe the various parts of an organism is protein; for example, hair, muscle, and skin are made up largely of protein. Other chemical compounds that make up the human body, such as carbohydrates, fats, and more-complex chemicals, are either synthesized by catalytic proteins (enzymes) or are deposited at specific times and in specific tissues under the influence of proteins. For example, the black-brown skin pigment melanin is synthesized by enzymes and deposited in special skin cells called melanocytes. Genes exert their effect mainly by determining the structure and function of the many thousands of different proteins, which in turn determine the characteristics of an organism. Generally, it is true to say that each protein is coded for by one gene, bearing in mind that the production of some proteins requires the cooperation of several genes.

Proteins are polymeric molecules; that is, they are made up of chains of monomeric elements, as is DNA. In proteins, the monomers are amino acids. Organisms generally contain 20 different types of amino acids, and the distinguishing factors that make one protein different from another are its length and specific amino acid sequence, which are determined by the number and sequence of nucleotide pairs in DNA. In other words, there is a colinearity (i.e., parallel structure) between the polymer that is DNA and the polymer that is protein.

Hence, genetic information flows from DNA into protein. However, this is not a single-step process. First, the nucleotide sequence of DNA is copied into the nucleotide sequence of single-stranded RNA in a process called transcription. Transcription of any one gene takes place at the chromosomal location of that gene. Whereas the unit of replication is a whole chromosome, the transcriptional unit is a relatively short segment of the chromosome, the gene. The active transcription of a gene depends on the need for the activity of that particular gene in a specific tissue or at a given time.

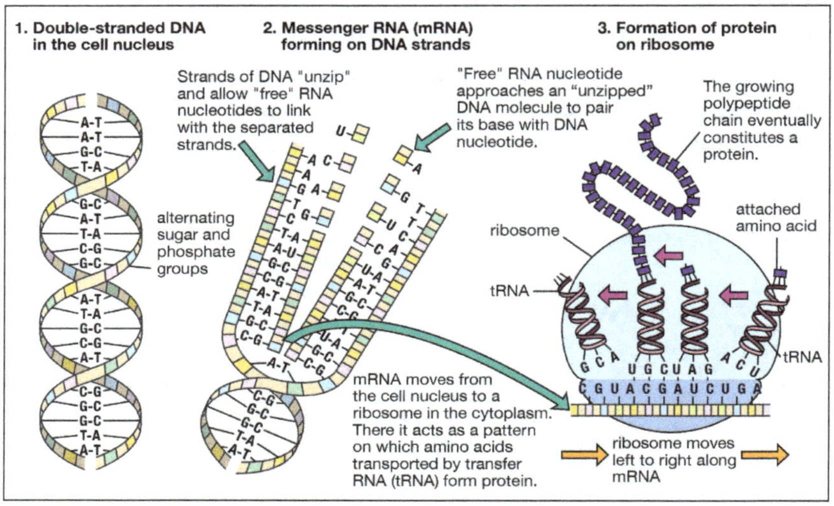

Figure indicates DNA in the cell nucleus carries a genetic code, which consists of sequences of adenine (A), thymine (T), guanine (G), and cytosine (C) . RNA, which contains uracil (U) instead of thymine, carries the code to protein-making sites in the cell. To make RNA, DNA pairs its bases with those of the "free" nucleotides. Messenger RNA (mRNA) then travels to the ribosomes in the cell cytoplasm, where protein synthesis occurs. The base triplets of transfer RNA (tRNA) pair with those of mRNA and at the same time deposit their amino acids on the growing protein chain. Finally, the synthesized protein is released to perform its task in the cell or elsewhere in the body.

The nucleotide sequence in RNA faithfully mirrors that of the DNA from which it was transcribed. The uracil in RNA has exactly the same hydrogen-bonding properties as thymine, so there are no changes at the information level. For most RNA molecules, the nucleotide sequence is converted into an amino acid sequence, a process called translation. In prokaryotes, translation begins during the transcription process, before the full RNA transcript is made. In eukaryotes, transcription finishes, and the RNA molecule passes from the nucleus into the cytoplasm, where translation takes place.

The genome of a type of virus called a retrovirus (of which the human immunodeficiency virus, or HIV, is an example) is composed of RNA instead of DNA. In a retrovirus, RNA is reverse transcribed into DNA, which can then integrate into the chromosomal DNA of the host cell that the retrovirus infects. The synthesis of DNA is catalyzed by the enzyme reverse transcriptase. The existence of reverse transcriptase shows that genetic information is capable of flowing from RNA to DNA in exceptional cases. Since it is believed that life arose in an RNA world, it is likely that the evolution of reverse transcriptase was an important step in the transition to the present DNA world.

Transcription

A gene is a functional region of a chromosome that is capable of making a transcript in response to appropriate regulatory signals. Therefore, a gene must not only be composed of the DNA sequence that is actually transcribed, but it must also include an adjacent regulatory, or control, region that is necessary for the transcript to be made in the correct developmental context.

The polymerization of ribonucleotides during transcription is catalyzed by the enzyme RNA polymerase. As with DNA replication, the two DNA strands must separate to expose the template. However, transcription differs from replication in that for any gene, only one of the DNA strands, the $3' \rightarrow 5'$ strand, is actually used as a template. Synthesis of RNA is in the $5' \rightarrow 3'$ direction, as with DNA. Hence, the growing point of the RNA chain is the $3'$ end, and polymerization is continuous as the RNA polymerase moves along the transcribed region. The RNA strand is extruded from the transcription complex like a tail, which grows longer as the transcription process advances. Eventually, a full-length transcript of RNA is produced, and this detaches from the DNA. The process is repeated, and multiple RNA transcripts are produced from one gene.

Prokaryotes possess only one type of RNA polymerase, but in eukaryotes there are several different types. RNA polymerase I synthesizes ribosomal RNA (rRNA), and RNA polymerase III synthesizes transfer RNA (tRNA) and other small RNAs. The types of RNA transcribed by these two polymerases are never translated into protein. RNA polymerase II transcribes the major type of genes, those genes that code for proteins.

Transcription of protein-coding genes results in a type of RNA called messenger RNA (mRNA), so named because it carries a genetic message from the gene on a nuclear chromosome into the cytoplasm, where it is acted upon by the protein-synthesizing apparatus. The transcription machinery contains many items in addition to the RNA polymerase. The successful binding of the RNA polymerase to the DNA "upstream" of the transcribed sequence depends upon the cooperation of many additional proteinaceous transcription factors. The region of the gene upstream from the region to be transcribed contains specific DNA sequences that are essential for the binding of transcription factors and a region called the promoter, to which the RNA polymerase binds. These sequences must be a specific distance from the transcriptional start site for successful operation. Various short base sequences in this regulatory region physically bind specific transcription factors by virtue of a lock-and-key fit between the DNA and the protein. As might be expected, a protein binds with the centre of the DNA molecule, which contains the sequence specificity, and not with the outside of the molecule, which is merely a uniform repetition of sugar and phosphate groups.

In eukaryotes, a key segment is the TATA box, a TATA sequence approximately 30 nucleotides upstream from the transcription start site. If this sequence is changed or moved, the rate of transcription drops drastically. The TATA box is bound by a transcription factor called the TATA-binding protein, which, together with RNA polymerase II and numerous other transcription factors, assembles in a precise sequence around the TATA box, binding to each other and to the DNA. Together, RNA polymerase and the transcription factors constitute the transcription complex.

The RNA polymerase is directed by the transcription complex to begin transcription at the proper site. It then moves along the template, synthesizing mRNA as it goes. At some position past the coding region, the transcription process stops. Bacteria have well-characterized specific termination sequences; however, in eukaryotes, termination signals are less well understood, and the transcription process stops at variable positions past the end of the coding sequence. A short nucleotide sequence downstream from the coding region acts as a signal for the RNA to be cut at that position, and this becomes the $3'$ end of the new RNA strand. Subsequently, approximately 200 adenine nucleotides are added to the $3'$ end to form what is called a poly(A) tail, which is characteristic of all eukaryotic DNA. At the $5'$ end of the mRNA, a modified guanine nucleotide, called a cap, is added. Noncoding nucleotide sequences called introns are excised from the RNA at this stage in a process called intron splicing. Molecular complexes called spliceosomes, which are composed of proteins and RNA, have RNA sequences that are complementary to the junction between introns and adjacent coding regions called exons. The intron is twisted into a loop and excised, and the exons are linked together. The resulting capped, tailed, and intron-free molecule is now mature mRNA.

Genetic Code

Hereditary information is contained in the nucleotide sequence of DNA in a kind of code. The coded information is copied faithfully into RNA and translated into chains of amino acids. Amino acid chains are folded into helices, zigzags, and other shapes and are sometimes associated with other amino acid chains. The specific amounts of amino acids in a protein and their sequence determine the protein's unique properties; for example, muscle protein and hair protein contain the same 20 amino acids, but the sequences of these amino acids in the two proteins are quite different. If the nucleotide sequence of mRNA is thought of as a written message, it can be said that this message is read by the translation apparatus in "words" of three nucleotides, starting at one end of the mRNA and proceeding along the length of the molecule. These three-letter words are called codons. Each codon stands for a specific amino acid, so if the message in mRNA is 900 nucleotides long, which corresponds to 300 codons, it will be translated into a chain of 300 amino acids.

Each of the three letters in a codon can be filled by any one of the four nucleotides; therefore, there are 4^3, or 64, possible codons. Each one of these 64 words in the codon dictionary has meaning. Most codons code for one of the 20 possible amino acids. Two amino acids, methionine and tryptophan, are each coded for by one codon only (AUG and UGG, respectively). The other 18 amino acids are coded for by two to six codons; for example, either of the codons UUU or UUC will cause the insertion of the amino acid phenylalanine into the growing amino acid chain. Three codons—UAG, UGA, and UAA—represent translation-termination signals and are called the stop codons. The first amino acid in an amino acid chain is methionine, encoded by an AUG codon. However, AUG codons are found throughout the coding sequence and are translated into methionines.

One of the surprising findings about the genetic codon dictionary is that, with a few exceptions, it is the same in all organisms. (One exception is mitochondrial DNA, which exhibits several differences from the standard genetic code and also between organisms.) The uniformity of the genetic code has been interpreted as an indication of the evolutionary relatedness of all organisms. For the purpose of genetic research, codon uniformity is convenient because any type of DNA can be translated in any organism.

Translation

The process of translation requires the interaction not only of large numbers of proteinaceous translational factors but also of specific membranes and organelles of the cell. In both prokaryotes and eukaryotes, translation takes place on cytoplasmic organelles called ribosomes. Ribosomes are aggregations of many different types of proteins and ribosomal RNA (rRNA). They can be thought of as cellular anvils on which the links of an amino acid chain are forged. A ribosome is a generic protein-making machine that can be recycled and used to synthesize many different types of proteins. A ribosome attaches to the 5′ end of the mRNA, begins translation at the start codon AUG, and translates the message one codon at a time until a stop codon is reached. Any one mRNA is translated many times by several ribosomes along its length, each one at a different stage of translation. In eukaryotes, ribosomes that produce proteins to be used in the same cell are not associated with membranes. However, proteins that must be exported to another location in the organism are synthesized on ribosomes located on the outside of flattened membranous chambers called the endoplasmic reticulum (ER). A completed amino acid chain is extruded into the inner cavity of the ER. Subsequently, the ER transports the proteins via small vesicles to another cytoplasmic organelle called the Golgi apparatus, which in turn buds off more vesicles that eventually fuse with the cell membrane. The protein is then released from the cell.

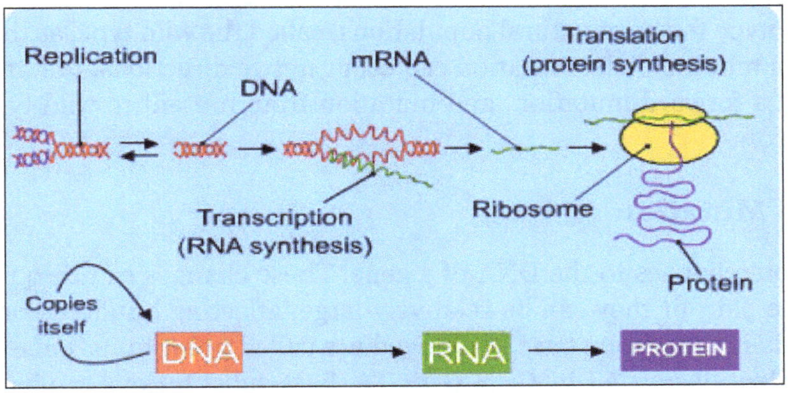

Synthesis of protein.

Another crucial component of the translational process is transfer RNA (tRNA). The function of any one tRNA molecule is to bind to a designated amino acid and carry it to a ribosome, where the amino acid is added to the growing amino acid chain. Each amino acid has its own set of tRNA molecules that will bind only to that specific amino acid. A tRNA molecule is a single nucleotide chain with several helical regions and a loop containing three unpaired nucleotides, called an anticodon. The anticodon of any one tRNA fits perfectly into the mRNA codon that codes for the amino acid attached to that tRNA; for example, the mRNA codon UUU, which codes for the amino acid phenylalanine, will be bound by the anticodon AAA. Thus, any mRNA codon that happens to be on the ribosome at any one time will solicit the binding only of the tRNA with the appropriate anticodon, which will align the correct amino acid for addition to the chain. A tRNA molecule and its attached amino acid must bind to the ribosome as well as to the codon during this amino acid chain-elongation process. A ribosome has two tRNA binding sites; at the first site, one tRNA attaches to the amino acid chain, and at the second site, another tRNA carrying the next amino acid is attached. After attachment, the first tRNA departs and recycles, whereas the second tRNA is now left holding the amino acid chain. At this time the ribosome moves to the next codon, and the whole process is successively repeated along the length of the mRNA until a stop codon is reached, at which time the completed amino acid chain is released from the ribosome.

The amino acid chain then spontaneously folds to generate the three-dimensional shape necessary for its function. Each amino acid has its own special shape and pattern of electrical charges on its surface, and ultimately these are what determine the overall shape of the protein. The protein's shape is stabilized by weak bonds that form between different parts of the chain. In some proteins, strong covalent bridges are formed between two cysteines at different sites in the chain. If the protein is composed of two or more amino acid chains, these also associate spontaneously and take on their most stable three-dimensional shape. For enzymes, shape determines the ability to bind to its specific substrate (i.e., the substance on which an enzyme acts). For structural proteins, the amino acid sequence determines whether it will be a filament, a sheet, a globule, or another shape.

Gene Mutation

Given the complexity of DNA and the vast number of cell divisions that take place within the lifetime of a multicellular organism, copying errors are likely to occur. If unrepaired, such errors will change the sequence of the DNA bases and alter the genetic code. Mutation is the random process whereby genes change from one allelic form to another. Scientists who study mutation use the most common genotype found in natural populations, called the wild type, as the standard against which to compare a mutant allele. Mutation can occur in two directions; mutation from wild type to mutant is called a forward mutation, and mutation from mutant to wild type is called a back mutation or reversion.

Mechanisms of Mutation

Mutations arise from changes to the DNA of a gene. These changes can be quite small, affecting only one nucleotide pair, or they can be relatively large, affecting hundreds or thousands of nucleotides. Mutations in which one base is changed are called point mutations—for example, substitution of the nucleotide pair AT by GC, CG, or TA. Base substitutions can have different consequences at the protein level. Some base substitutions are "silent," meaning that they result in a new codon that codes for the same amino acid as the wild type codon at that position or a codon that codes for a different amino acid that happens to have the same properties as those in the wild type. Substitutions that result in a functionally different amino acid are called "missense" mutations; these can lead to alteration or loss of protein function. A more severe type of base substitution, called a "nonsense" mutation, results in a stop codon in a position where there was not one before, which causes the premature termination of protein synthesis and, more than likely, a complete loss of function in the finished protein.

Another type of point mutation that can lead to drastic loss of function is a frameshift mutation, the addition or deletion of one or more DNA bases. In a protein-coding gene, the sequence of codons starting with AUG and ending with a termination codon is called the reading frame. If a nucleotide pair is added to or subtracted from this sequence, the reading frame from that point will be shifted by one nucleotide pair, and all of the codons downstream will be altered. The result will be a protein whose first section (before the mutational site) is that of the wild type amino acid sequence, followed by a tail of functionally meaningless amino acids. Large deletions of many codons will not only remove amino acids from a protein but may also result in a frameshift mutation if the number of nucleotides deleted is not a multiple of three. Likewise, an insertion of a block of nucleotides will add amino acids to a protein and perhaps also have a frameshift effect.

A number of human diseases are caused by the expansion of a trinucleotide pair repeat. For example, fragile-X syndrome, the most common type of inherited mental retardation in humans, is caused by the repetition of up to 1,000 copies of a CGG repeat in a gene on the X chromosome.

The impact of a mutation depends upon the type of cell involved. In a haploid cell, any mutant allele will most likely be expressed in the phenotype of that cell. In a diploid cell, a dominant mutation will be expressed over the wild type allele, but a recessive mutation will remain masked by the wild type. If recessive mutations occur in both members of one gene pair in the same cell, the mutant phenotype will be expressed. Mutations in germinal cells (i.e., reproductive cells) may be passed on to successive generations. However, mutations in somatic (body) cells will exert their effect only on that individual and will not be passed on to progeny.

The impact of an expressed somatic mutation depends upon which gene has been mutated. In most cases, the somatic cell with the mutation will die, an event that is generally of little consequence in a multicellular organism. However, mutations in a special class of genes called proto-oncogenes can cause uncontrolled division of that cell, resulting in a group of cells that constitutes a cancerous tumour.

Mutations can affect gene function in several different ways. First, the structure and function of the protein coded by that gene can be affected. For example, enzymes are particularly susceptible to mutations that affect the amino acid sequence at their active site (i.e., the region that allows the enzyme to bind with its specific substrate). This may lead to enzyme inactivity; a protein is made, but it has no enzymatic function. Second, some nonsense or frameshift mutations can lead to the complete absence of a protein. Third, changes to the promoter region of the gene can result in gene malfunction by interfering with transcription. In this situation, protein production is either inhibited or it occurs at an inappropriate time because of alterations somewhere in the regulatory region. Fourth, mutations within introns that affect the specific nucleotide sequences that direct intron splicing may result in an mRNA that still contains an intron. When translated, this extra RNA will almost certainly be meaningless at the protein level, and its extra length will lead to a functionless protein. Any mutation that results in a lack of function for a particular gene is called a "null" mutation. Less-severe mutations are called "leaky" mutations because some normal function still "leaks through" into the phenotype.

Most mutations occur spontaneously and have no known cause. The synthesis of DNA is a cooperative venture of many different interacting cellular components, and occasionally mistakes occur that result in mutations. Like many chemical structures, the bases of DNA are able to exist in several conformations called isomers. The keto form of a DNA base is the normal form that gives the molecule its standard base-pairing properties. However, the keto form occasionally changes spontaneously to the enol form, which has different base-pairing properties. For example, the keto form of cytosine pairs with guanine (its normal pairing partner), but the enol form of cytosine pairs with adenine. During DNA replication, this adenine base will act as the template for thymine in the newly synthesized strand. Therefore, a CG base pair will have mutated to a TA base pair. If this change results in a functionally different amino acid, then a missense mutation may result. Another spontaneous event that can lead to mutation is depurination, the complete loss of a purine base (adenine or guanine) at some location in the DNA. The resulting gap can be filled by any base during subsequent replications.

Researchers have demonstrated that ionizing radiation, some chemicals, and certain viruses are capable of acting as mutagens—agents that can increase the rate at which mutations occur. Some mutagens have been implicated as a cause of cancer. For example, ultraviolet (UV) radiation from the sun is known to cause skin cancer, and cigarette smoke is a primary cause of lung cancer.

Repair of Mutation

A variety of mechanisms exists for repairing copying errors caused by DNA damage. One of the best-studied systems is the repair mechanism for damage caused by ultraviolet radiation. Ultraviolet radiation joins adjacent thymines, creating thymine dimers, which, if not repaired, may cause mutations. Special repair enzymes either cut the bond between the thymines or excise the bonded dimer and replace it with two single thymines. If both of these repair methods fail, a third method allows the DNA replication process to bypass the dimer; however, it is this bypass system that causes most mutations because bases are then inserted at random opposite the thymine dimer. Xeroderma pigmentosum, a severe hereditary disease of humans, is caused by a mutation in a gene coding for one of the thymine dimer repair enzymes. Individuals with this disease are highly susceptible to skin cancer.

Reverse mutation from the aberrant state of a gene back to its normal, or wild type, state can result in a number of possible molecular changes at the protein level. True reversion is the reversal of the original nucleotide change. However, phenotypic reversion can result from changes that restore a different amino acid with properties identical to the original. Second-site changes within a protein can also restore normal function. For example, an amino acid change at a site different from that altered by the original mutation can sometimes interact with the amino acid at the first mutant site to restore a normal protein shape. Also, second-site mutations at other genes can act as suppressors, restoring wild type function. For example, mutations in the anticodon region of a tRNA gene can result in a tRNA that sometimes inserts an amino acid at an erroneous stop codon; if the original mutation is caused by a stop codon, which arrests translation at that point, then a tRNA anticodon change can insert an amino acid and allow translation to continue normally to the end of the mRNA. Alternatively, some mutations at separate genes open up a new biochemical pathway that circumvents the block of function caused by the original mutation.

Regulation of Gene Expression

Not all genes in a cell are active in protein production at any given time. Gene action can be switched on or off in response to the cell's stage of development and external environment. In multicellular organisms, different kinds of cells express different parts of the genome. In other words, a skin cell and a muscle cell contain exactly the same genes, but the differences in structure and function of these cells result from the selective expression and repression of certain genes.

In prokaryotes and eukaryotes, most gene-control systems are positive, meaning that a gene will not be transcribed unless it is activated by a regulatory protein. However, some bacterial genes show negative control. In this case the gene is transcribed continuously unless it is switched off by a regulatory protein. An example of negative control in prokaryotes involves three adjacent genes used in the metabolism of the sugar lactose by E. coli. The part of the chromosome containing the genes concerned is divided into two regions, one that includes the structural genes (i.e., those genes that together code for protein structure) and another that is a regulatory region. This overall unit is called an operon. If lactose is not present in a cell, transcription of the genes that code for the lactose-processing enzymes—β-galactosidase, permease, and transacetylase—is turned off. This is achieved by a protein called the lac repressor, which is produced by the repressor gene and binds to a region of the operon called the operator. Such binding prevents RNA polymerase, which initially binds at the adjacent promoter, from moving into the coding region. If lactose enters the cell, it binds to the lac repressor and induces a change of shape in the repressor so that it can no longer bind to the DNA at the operon. Consequently, the RNA polymerase is able to travel from the promoter down the three adjacent protein-coding regions, making one continuous transcript. This three-gene transcript is subsequently translated into three separate proteins.

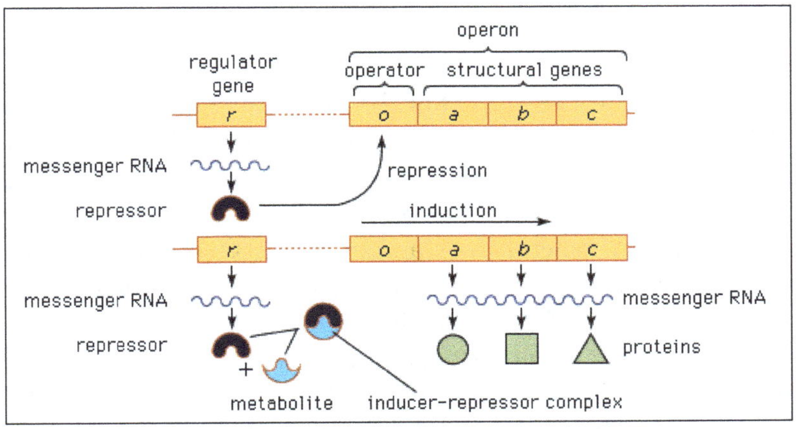

Model of the operon and its relation to the regulator gene.

Although the operon model has proved a useful model of gene regulation in bacteria, different regulatory mechanisms are employed in eukaryotes. First, there are no operons in eukaryotes, and each gene is regulated independently. Furthermore, the series of events associated with gene expression in higher organisms is much more complex than in prokaryotes and involves multiple levels of regulation.

In order for a gene to produce a functional protein, a complex series of steps must occur. Some type of signal must initiate the transcription of the appropriate region along the DNA, and, finally, an active protein must be made and sent to the appropriate location to perform its specific task. Regulation can be exerted at many different places along this pathway. The fundamental level of control is the rate of transcription. Transcription itself is also a complex process with many different components, and each one is a potential point of control. Regulatory proteins called activators or enhancers are needed for the transcription of genes at a specific time or in a certain cell. Thus, control is positive (not negative as in the lac operon) in that these proteins are necessary for the promotion of transcription. Activators bind to specific regions of the DNA in the upstream regulatory region, some very distant from the binding of the initiation complex.

Following the transcription of DNA into RNA, a process of editing and splicing takes place in which noncoding nucleotide sequences called introns are excised from the primary transcript, resulting in functional mRNA. For most genes this is a routine step in the production of mRNA, but in some genes there are alternative ways to splice the primary transcript, resulting in different mRNAs, which in turn result in different proteins.

Some genes are controlled at the translational and post-translational levels. One type of translational control is the storage of uncapped mRNA to meet future demands for protein synthesis. In other cases, control is exerted through the stability or instability of mRNA. The rate of translation of some mRNAs can also be regulated. Post-translationally, certain proteins (e.g., insulin) are synthesized in an inactive form and must be chemically modified to become active. Other proteins are targeted to specific locations inside the cell (e.g., mitochondria) by means of highly specific amino acid sequences at their ends, called leader sequences; when the protein reaches its correct site, the leader segment is cut off, and the protein begins to function. Post-translational control is also exerted through mRNA and protein degradation.

Repetitive DNA

One major difference between the genomes of prokaryotes and eukaryotes is that most eukaryotes contain repetitive DNA, with the repeats either clustered or spread out between the unique genes. There are several categories of repetitive DNA: (1) single copy DNA, which contains the structural genes (protein-coding sequences), (2) families of DNA, in which one gene somehow copies itself, and the repeats are located in small clusters (tandem repeats) or spread throughout the genome (dispersed repeats), and (3) satellite DNA, which contains short nucleotide sequences repeated as many as thousands of times. Such repeats are often found clustered in tandem near the centromeres (i.e., the attachment points for the nuclear spindle fibres that move chromosomes during cell division). Microsatellite DNA is composed of tandem repeats of two nucleotide pairs that are dispersed throughout the genome. Minisatellite DNA, sometimes called variable number tandem repeats (VNTRs), is composed of blocks of longer repeats also dispersed throughout the genome. There is no known function for satellite DNA, nor is it known how the repeats are created. There is a special class of relatively large DNA elements called transposons, which can make replicas of themselves that "jump" to different locations in the genome; most transposons eventually become inactive and no longer move, but, nevertheless, their presence contributes to repetitive DNA.

Extranuclear DNA

All of the genetic information in a cell was initially thought to be confined to the DNA in the chromosomes of the cell nucleus. It is now known that small circular chromosomes, called extranuclear, or cytoplasmic, DNA, are located in two types of organelles found in the cytoplasm of the cell. These organelles are the mitochondria in animal and plant cells and the chloroplasts in plant cells. Chloroplast DNA (cpDNA) contains genes that are involved with aspects of photosynthesis and with components of the special protein-synthesizing apparatus that is active within the organelle. Mitochondrial DNA (mtDNA) contains some of the genes that participate in the conversion of the energy of chemical bonds into the energy currency of the cell—a chemical called adenosine triphosphate (ATP)—as well as genes for mitochondrial protein synthesis.

The cells of several groups of organisms contain small extra DNA molecules called plasmids. Bacterial plasmids are circular DNA molecules; some carry genes for resistance to various agents in the environment that would be toxic to the bacteria (e.g., antibiotics). Many fungi and some plants possess plasmids in their mitochondria; most of these are linear DNA molecules carrying genes that seem to be relevant only to the propagation of the plasmid and not the host cell.

Heredity and Evolution

At the centre of the theory of evolution as proposed by Charles Darwin and Alfred Russell Wallace were the concepts of variation and natural selection. Hereditary variants were thought to arise naturally in populations, and then these were either selected for or against by the contemporary environmental conditions. In this way, subsequent generations either became enriched or impoverished for specific variant types. Over the long term, the accumulation of such changes in populations could lead to the formation of new species and higher taxonomic categories. However, although hereditary change was basic to the theory, in the 19th-century world of Darwin and Wallace, the fundamental unit of heredity—the gene—was unknown. The birth and proliferation of the science of genetics in the 20th century after the discovery of Mendel's laws made it possible to consider the process of evolution by natural selection in terms of known genetic processes.

Population Genetics

Because the processes of variation and selection take place at the population level, the basic theory of the genetics of evolutionary change is contained in the general area known as population genetics.

A simple way of viewing evolutionary change at the genotypic level would be to invent some hypothetical ancestral genotype, such as AAbbccDDEE, and an "evolved" derivative, such as aaBBccddee. (For illustrative purposes, only five genes are used, and these are assumed to be all homozygous.) Also, for simplicity it can be assumed that in both the ancestral and the evolved populations all individuals are identical. Clearly for all the genes except cc, a new allele completely replaces the original allele, and the new alleles can be either dominant or recessive. For example, in the case of the first gene, in the ancestral population all alleles are A, and in the evolved population all are a. For a to replace A, the population must go through stages in which there are mixtures of A and a alleles present in the population at the same time. In population genetics, allele frequency is the measurement of the commonness of an allele. The convention is to let the frequency of a dominant allele be p and that of a recessive allele q. Both are generally expressed as decimal fractions. In the above example, p changes from 1 to 0, and q changes from 0 to 1. Since there are only two alleles in this example, p + q must always equal 1. In the intermediate stages, there must be times when there are intermediate allele frequencies, for example when p = 0.4 and q = 0.6.

What can be said about genotype frequencies in the intermediate populations? In the ancestral and derived populations there must have been the following genotypic frequencies:

$$\text{Ancestral } AA = 1,\ Aa = 0,\ aa = 0$$

$$\text{Evolved } AA = 0,\ Aa = 0,\ aa = 1$$

Intuitively it seems that, in the intermediate stages, there must be more-complex proportions, including some heterozygotes. One possible intermediate stage is as follows:

$$AA = 0.30,\ Aa = 0.20,\ aa = 0.50$$

The allele frequencies at such an intermediate stage can be calculated by "adding up" the alleles. Hence, the frequency of A will be 0.30 plus $^1/_2$ of 0.20 because the heterozygotes only carry one A allele. This is written:

$$p = 0.30 + {}^{0.20}/_2 = 0.40$$

Similarly, $q = 0.50 + {}^{0.20}/_2 = 0.60$

(Noting these values for p and q, it is possible that this could have been the population discussed earlier, in which these specific values for p and q were hypothesized.)

In general, if D = frequency of homozygous dominants, R = frequency of homozygous recessives, and H = frequency of heterozygotes then:

$$p = D + {}^{H}/_{2}$$

and,

$$q = R + {}^{H}/_{2}$$

Allele frequency is the simpler descriptor, and it forms the central tool of population genetics. Hence, the genetic basis of evolutionary change at the population level is described in terms of changes of allele frequencies.

Hardy-Weinberg Equilibrium

It is a curious fact that populations show no inherent tendency to change allele or genotype frequencies. In the absence of selection or any of the other forces that can drive evolution, a population with given values of p and q will settle into a special stable set of genotypic proportions called a Hardy-Weinberg equilibrium. This principle was first realized by Godfrey Harold Hardy and Wilhelm Weinberg in 1908. The Hardy-Weinberg equilibrium of a population with allele frequencies p and q is defined by the set of genotypic frequencies p^2 of AA, 2pq of Aa, and q^2 of aa.

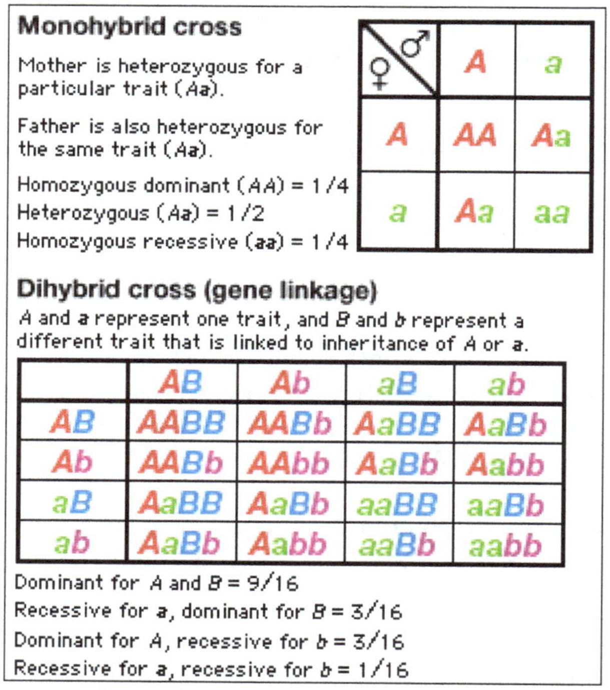

Punnett square diagrams are used to predict all the possible gene combinations that could result from the mating of parents with known genotypes for a particular trait. A monohybrid cross represents the inheritance pattern of a single trait, whereas a dihybrid cross represents the inheritance patterns of two traits that are linked.

When such a population reproduces itself to make a new generation, the lack of change is made apparent. It is intuitive that the allele frequencies p and q in the population are also measures of the frequencies of eggs and sperm used in creating a new generation. The new generation produced from the zygotes has exactly the same genotypic proportions as the first generation (the parents of the zygote).

Some specific allele frequencies, 0.7 for p and 0.3 for q, can be used to illustrate the calculation of the genotypic frequencies that constitute the Hardy-Weinberg equilibrium:

$$p \times p = 0.7 \times 0.7 = 0.49 \text{ of AA}$$

$$2 \times p \times q = 2 \times 0.7 \times 0.3 = 0.42 \text{ of Aa}$$

$$q \times q = 0.3 \times 0.3 = 0.09 \text{ of aa}$$

When this population reproduces, there will be $0.49 + 0.21 = 0.7$ of A gametes and $0.09 + 0.21 = 0.3$ of a gametes, and, when these gametes combine, the population in the next generation will clearly have the same genotypic proportions as the previous one.

These simple calculations rely on several underlying assumptions. Perhaps the most crucial one is that there is random mating, or mating regardless of the genotype of the partner. In addition, the population must be large, and there can be no other pressures, such as selection, that can change allele frequencies. Despite these stringent requirements, many natural populations that have been studied are in Hardy-Weinberg equilibrium for the genes under investigation. The Hardy-Weinberg equilibrium constitutes an important benchmark for population genetic analysis.

If the Hardy-Weinberg principle of population genetics shows that there is no inherent tendency for evolutionary change, then how does change occur? This is considered in the following sections.

Changes in Gene Frequencies

Selection

One assumption behind the calculation of unchanging genotypic frequencies in Hardy-Weinberg equilibrium is that all genotypes have the same fitness. In genetics, fitness does not necessarily have to do with muscles; fitness is a measure of the ability to produce fertile offspring. In reality, the fitnesses of different genotypes are highly variable. The genotype with the greatest fitness is given the fitness value (w) of 1, and the lesser fitnesses are fractions of 1. For example, if snails of genotypes AA and Aa were to have an average of 100 offspring but those of genotype aa only 70, then the fitnesses of these three genotypes would be 1, 1, and 0.7, respectively. The proportional difference from the most fit is called the selection coefficient, s. Hence, $s = 1 - w$.

Alleles carried by less-fit individuals will be gradually lost from the population, and the relevant allele frequency will decline. This is the fundamental way in which natural selection operates in a population. Selection against dominant alleles is relatively efficient, because these are by definition expressed in the phenotype. Selection against recessive alleles is less efficient, because these alleles are sheltered in heterozygotes. Even though populations under selection technically are not in Hardy-Weinberg equilibrium, the proportions of the formula can be used as an approximation to show the relative proportions of homozygous recessives and heterozygotes. If a rare deleterious recessive allele is of frequency $^1/_{50}$ in the population, then $(^1/_{50})^2$, or 1 out of 2,500, individuals will express the recessive phenotype and be a candidate for negative selection. Heterozygotes will be at a frequency of $2pq = 2 \times {}^{49}/_{50} \times {}^1/_{50}$, or about 1 in 25. In other words, the heterozygotes are 100 times more common than recessive homozygotes; hence, most of the recessive alleles in a population will escape selection.

Because of the sheltering effect of heterozygotes, selection against recessive phenotypes changes the frequency of the recessive allele slowly. Even if the most severe level of selection is imposed, giving the recessive phenotype a fitness of zero (no fertile offspring), the recessive allele frequency (expressed as a fraction of the form $^1/_x$) will increase in denominator by 1 in every generation. Therefore, to halve an allele frequency from $^1/_{50}$ to $^1/_{100}$ would proceed slowly from $^1/_{50}$ to $^1/_{51}$, $^1/_{52}$, $^1/_{53}$, and so on and would take 50 generations to get to $^1/_{100}$. For lower intensities of selection, the progress would be even slower.

A different type of natural selection occurs when the fitness of a heterozygote exceeds the fitness of both homozygotes. The maintenance in human populations of the severe hereditary disease sickle cell anemia is owing to this form of selection. The disease allele (Hb^S) produces a specific type of hemoglobin that causes distortion (sickling) of the red blood cells in which the hemoglobin is carried. (Normal hemoglobin is coded by another allele, Hb^A). Accordingly, the possible genotypes are Hb^AHb^A, Hb^AHb^S, and Hb^SHb^S. The latter individuals are homozygous for the sickle cell allele and will develop severe anemia because the oxygen transporting property of their blood is compromised. While the condition is not lethal before birth, such individuals rarely survive long enough to reproduce. On these grounds it might be expected that the disease allele would be selected against, driving the allele frequency to very low levels. However, in tropical areas of the world, the allele and the disease are common. The explanation is that the Hb^AHb^S heterozygote is fitter and capable of leaving more offspring than is the homozygous normal Hb^AHb^A in an environment containing the falciparum form of malaria. This extra measure of protection is evidently provided by the sickle cell hemoglobin, which is detrimental to the malaria parasite. In malarial environments, therefore, populations that contain the sickle cell gene have advantages over populations free of this gene. The former populations are in less danger from malaria, although they "pay" for this advantage by sacrificing in every generation some individuals who die of anemia.

Mutation

Genetics has shown that mutation is the ultimate source of all hereditary variation. At the level of a single gene whose normal functional allele is A, it is known that mutation can change it to a nonfunctional recessive form, a. Such "forward mutation" is more frequent than "back mutation" (reversion), which converts a into A. Molecular analysis of specific examples of mutant recessive alleles has shown that they are generally a heterogeneous set of small structural changes in the DNA, located throughout the segment of DNA that constitutes that gene. Hence, in an example from medical genetics, the disease phenylketonuria is inherited as a recessive phenotype and is ascribed to a causative allele that generally can be called k. However, sequencing alleles of many independent cases of phenylketonuria has shown that this k allele is in fact a set of many different kinds of mutational changes, which can be in any of the protein-coding regions of that gene.

Recessive deleterious mutations are relatively rare, generally in the order of 1 per 105 or 106 mutant gametes per generation. Their constant occurrence over the generations, combined with the even greater rarity of back mutations, leads to a gradual accumulation in the population. This accumulation process is called mutational pressure.

Effect of nucleotide substitutions on codons for amino acids

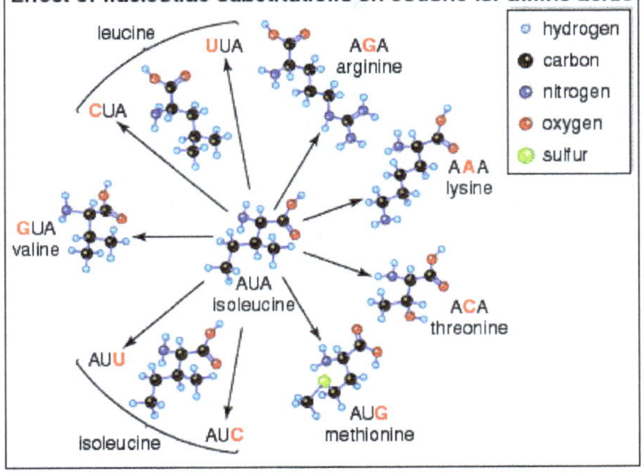

The above figure shows effect of base substitutions, or point mutations, on the messenger-RNA codon AUA, which codes for the amino acid isoleucine. Substitutions (red letters) at the first, second, or third position in the codon can result in nine new codons corresponding to six different amino acids in addition to isoleucine itself. The chemical properties of some of these amino acids are quite different from those of isoleucine. Replacement of one amino acid in a protein by another can seriously affect the protein's biological function.

Since mutational pressure to a deleterious recessive allele and selection pressure against the homozygous recessives are forces that act in opposite directions, another type of equilibrium is attained that effectively sets the value of q. Mathematically, q is determined by the following expression in which u is the net mutation rate of A to a, and s is the selection coefficient presented above:

$$q^2 = ({}^{u}/_{s}), \text{ or } q = \sqrt{(u/s)}$$

Nonrandom Mating

Many species engage in alternatives to random mating as normal parts of their cycle of sexual reproduction. An important exception is sexual selection, in which an individual chooses a mate on the basis of some aspect of the mate's phenotype. The selection can be based on some display feature such as bright feathers, or it may be a simple preference for a phenotype identical to the individual's own (positive assortative mating).

Two other important exceptions are inbreeding (mating with relatives) and enforced outbreeding. Both can shift the equilibrium proportions expected under Hardy-Weinberg calculations. For example, inbreeding increases the proportions of homozygotes, and the most extreme form of inbreeding, self-fertilization, eventually eliminates all heterozygotes.

Inbreeding and outbreeding are evolutionary strategies adopted by plants and animals living under certain conditions. Outbreeding brings gametes of different genotypes together, and the resulting individual differs from the parents. Increased levels of variation provide more evolutionary flexibility. All the showy colors and shapes of flowers are to promote this kind of exchange. In contrast, inbreeding maintains uniform genotypes, a strategy successful in stable ecological habitats.

In humans, various degrees of inbreeding have been practiced in different cultures. In most cultures today, matings of first cousins are the maximal form of inbreeding condoned by society. Apart from ethical considerations, a negative outcome of inbreeding is that it increases the likelihood of homozygosity of deleterious recessive alleles originating from common ancestors, called homozygosity by descent. The inbreeding coefficient F is a measure of the likelihood of homozygosity by descent; for example, in first-cousin marriages, $F={}^{1}/_{16}$. A large proportion of recessive hereditary diseases can be traced to first-cousin marriages and other types of inbreeding.

Random Genetic Drift

In populations of finite size, the genetic structure of a new generation is not necessarily that of the previous one. The explanation lies in a sampling effect, based on the fact that a subsample from any large set is not always representative of the larger set. The gametes that form any generation can be thought of as a sample of the alleles from the parental one. By chance the sample might not be random; it could be skewed in either direction. For example, if $p = 0.600$ and $q = 0.400$, sampling "error" might result in the gametes having a p value of 0.601 and a q of 0.399. If by chance this skewed sampling occurs in the same direction from generation to generation, the allele frequency can change radically. This process is known as random genetic drift. As might be expected, the smaller the population, the greater chance of sampling error and hence significant levels of drift in any one generation. In extreme cases, drift over the generations can result in the complete loss of one allele; in these occurrences the other is said to be fixed.

Other cases of sampling error occur when new colonies of plants or animals are founded by small numbers of migrants (founder effect) and when there is radical reduction in population size because of a natural catastrophe (population bottleneck). One inevitable effect of these processes is a reduction in the amount of variation in the population after the size reduction. Two species that have gone through drastic bottlenecks with the associated reduction of genetic variation are cheetahs (Africa) and northern elephant seals (North America).

Microevolution

There is ample evidence that the processes are at work in natural populations. Together, these changes are called microevolution—in other words, small-scale evolution. Even within the relatively short period of time since Darwin, it has been possible to document such processes. Allelic variation has been found to be common in nature. It is detected as polymorphism, the presence of two or more distinct hereditary forms associated with a gene. Polymorphism can be morphological, such as blue and brown forms of a species of marine mussel, or molecular, detectable only at the DNA or protein level. Although much of this polymorphism is not understood, there are enough examples of selection of polymorphic forms to indicate that it is potentially adaptive. Selection has been observed favouring melanic (dark) forms of peppered moths in industrial areas and favouring resistance to toxic agents such as the insecticide DDT, the rat poison warfarin, and the virus that causes the disease myxomatosis in rabbits.

Industrial melanism in the peppered moth.

The light gray form of the peppered moth (Biston betularia) is inconspicuous on the lichen-covered tree trunk. In contrast, the dark (melanic) morph stands out, leaving it vulnerable to predation by birds. The gradual darkening of the wings of the melanic peppered moth, the result of living in woodlands darkened by industrial pollution, is an example of industrial melanism.

More-complex genetic changes have been documented, leading to special locally adapted "ecotypes." Anoles (a type of lizard) on certain Caribbean islands show convincing examples of adaptations to specific habitats, such as tree trunks, tree branches, or grass. Introductions of lizards onto uncolonized islands result in demonstrable microevolutionary adaptations to the various vacant niches. On the Galapagos Islands, studies over several decades have documented adaptive changes in the beaks of finches. In some studies, documented changes have led to incipient new species. An example is the apple maggot, the larva of a fly in North America that has evolved from a similar fly living on hawthorns—all in the period since the introduction of apples. The formation of new species was a key component of Darwin's original theory. Now it appears that the accumulation of enough small-scale genetic changes can lead to the inability to mate with members of an ancestral population; such reproductive isolation is the key step in species formation.

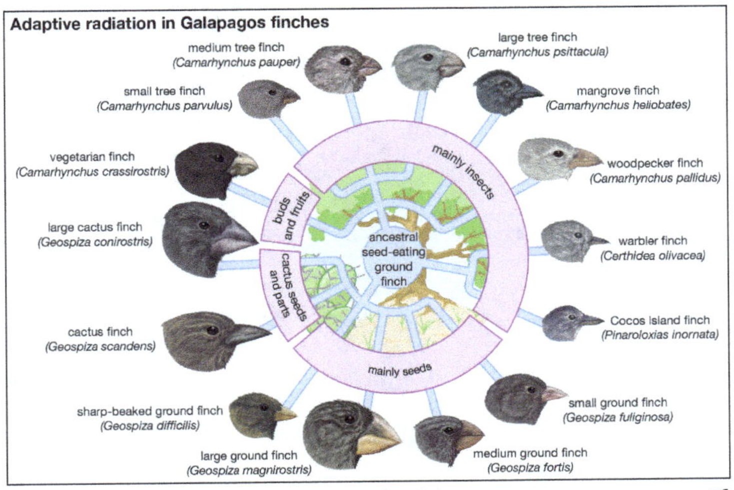

Fourteen species of Galapagos finches that evolved from a common ancestor. The different shapes of their bills, suited to different diets and habitats, show the process of adaptive radiation.

It is reasonable to assume that the continuation of microevolutionary genetic changes over very long periods of time can give rise to new major taxonomic groups, the process of macroevolution. There are few data that bear directly on the processes of macroevolution, but gene analysis does provide a way for charting macroevolutionary relationships indirectly.

DNA Phylogeny

The ability to isolate and sequence specific genes and genomes has been of great significance in deducing trees of evolutionary relatedness. An important discovery that enables this sort of analysis is the considerable evolutionary conservation between organisms at the genetic level. This means that different organisms have a large proportion of their genes in common, particularly those that code for proteins at the central core of the chemical machinery of the cell. For example, most organisms have a gene coding for the energy-producing protein cytochrome C, and furthermore, this gene has a very similar nucleotide sequence in all organisms (that is, the sequence is conserved). However, the sequences of cytochrome C in different organisms do show differences, and the key to phylogeny is that the differences are proportionately fewer between organisms that are closely related. The interpretation of this observation is that organisms that share a common ancestor also share common DNA sequences derived from that ancestor. When one ancestral species splits into two, differences accumulate as a result of mutations, a process called divergence. The greater the amount of divergence, the longer must have been the time since the split occurred. To carry out this sort of analysis, the DNA sequence data are fed into a computer. The computer positions similar species together on short adjacent branches showing a relatively recent split and dissimilar species on long branches from an ancient split. In this way a molecular phylogenetic tree of any number of organisms can be drawn.

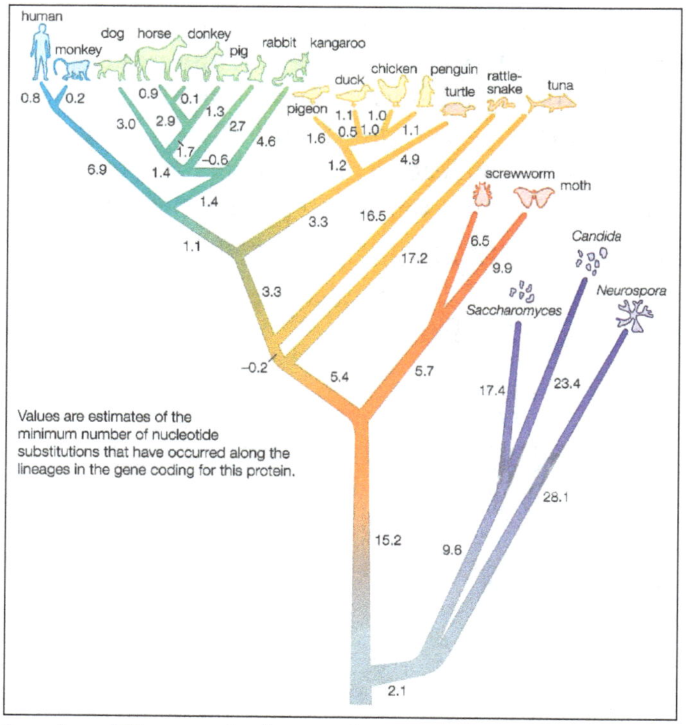

Phylogeny based on differences in the protein sequence of cytochrome c in organisms ranging from Neurospora mold to humans.

DNA difference in some cases can be correlated with absolute dates of divergence as deduced from the fossil record. Then it is possible to calculate divergence as a rate. It has been found that divergence is relatively constant in rate, giving rise to the idea that there is a type of "molecular clock" ticking in the course of evolution. Some ticks of this clock (in the form of mutations) are significant in terms of adaptive changes to the gene, but many are undoubtedly neutral, with no significant effect on fitness.

One of the interesting discoveries to emerge from molecular phylogeny is that gene duplication has been common during evolution. If an extra copy of a gene can be made, initially by some cellular accident, then the "spare" copy is free to mutate and evolve into a separate function.

Molecular phylogeny of some genes has also pointed to unexpected cases of, say, a plant gene nested within a tree of animal genes of that type or a bacterial gene nested within a plant phylogenetic tree. The explanation for such anomalies is that there has been horizontal transmission from one group to another. In other words, on rare occasions a gene can hop laterally from one species to another. Although the mechanisms for horizontal transmission are presently not known, one possibility is that bacteria or viruses act as natural vectors for transferring genes.

Synteny

Genomic sequencing and mapping have enabled comparison of the general structures of genomes of many different species. The general finding is that organisms of relatively recent divergence show similar blocks of genes in the same relative positions in the genome. This situation is called synteny, translated roughly as possessing common chromosome sequences. For example, many of the genes of humans are syntenic with those of other mammals—not only apes but also cows, mice, and so on. Study of synteny can show how the genome is cut and pasted in the course of evolution.

Polyploidy

Genomic analysis also has shown that one of the important mechanisms of evolution is multiplication of chromosome sets, resulting in polyploidy ("many genomes"). In plants and animals, spontaneous doubling of chromosomes can occur. In some plants, the chromosomes of two related species unite via cross-pollination to form a fusion product. This product is sterile because each chromosome needs a pairing partner in order for the plant to be fertile. However, the chromosomes of the fusion product can accidentally double, resulting in a new, fertile species. Wheat is an example of a plant that evolved by this means through a union between wild grasses, but a large proportion of plants went through similar ancestral polyploidization.

Human Evolution

Many of the techniques of evolutionary genetics can be applied to the evolution of humans. Charles Darwin created a large controversy in Victorian England by suggesting in his book The Descent of Man that humans and apes share a common ancestor. Darwin's assertion was based on the many shared anatomical features of apes and humans. DNA analysis has supported this hypothesis. At the DNA sequence level, the genomes of humans and chimpanzees are 99 percent identical. Furthermore, when phylogenetic trees are constructed using individual genes, humans and apes cluster together in short terminal branches of the trees, suggesting very recent divergence. Synteny too is impressive, with relatively minor chromosomal rearrangements.

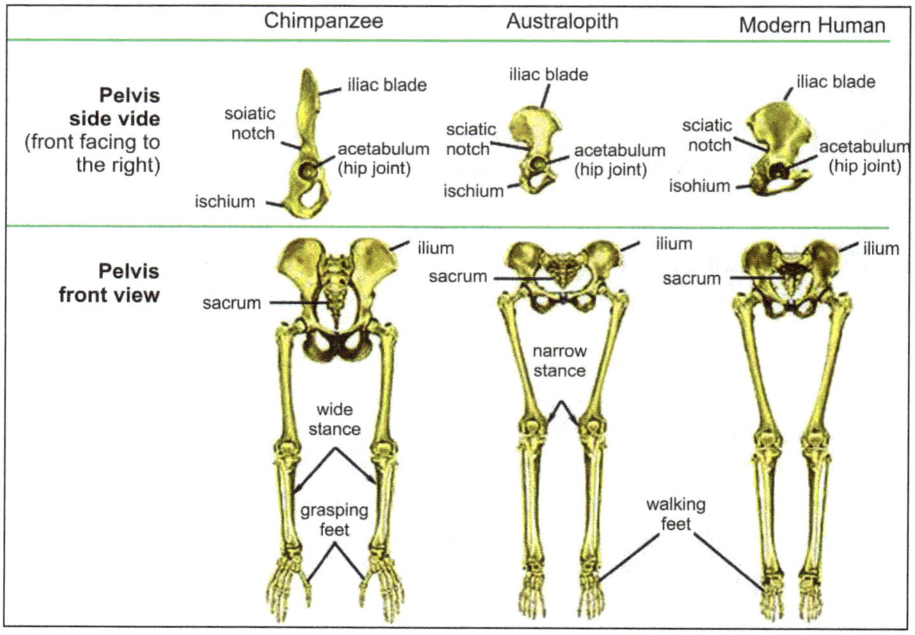

Comparison of the pelvis and lower limbs of a chimpanzee, an australopith, and a modern human.

Fossils have been found of various extinct forms considered to be intermediates between apes and humans. Notable is the African genus Australopithecus, generally believed to be one of the earliest hominins and an intermediate on the path of human evolution. The first toolmaker was Homo habilis, followed by Homo erectus and finally Homo sapiens (modern humans). H. habilis fossils have been found only in Africa, whereas fossils of H. erectus and H. sapiens are found throughout the Old World. Phylogenetic trees based on DNA sequencing of all peoples have shown that Africans represent the root of the trees. This is interpreted as evidence that H. sapiens evolved in Africa, spread throughout the globe, and outcompeted H. erectus wherever the two cohabited.

Variations of DNA, either unique alleles of individual genes or larger-sized blocks of variable structure, have been used as markers to trace human migrations across the globe. Hence, it has been possible to trace the movement of H. sapiens out of Africa and into Europe and Asia and, more recently, to the American continents. Also, genetic markers are useful in plotting human migrations that occurred in historical time. For example, the invasion of Europe by various Asian conquerors can be followed using blood-type alleles.

As humans colonized and settled permanently in various parts of the world, they differentiated themselves into distinct groups called races. Undoubtedly, many of the features that distinguish races, such as skin colour or body shape, were adaptive in the local settings, although such adaptiveness is difficult to demonstrate. Nevertheless, genomic analysis has revealed that the concept of race has little meaning at the genetic level. The differences between races are superficial, based on the alleles of a relatively small number of genes that affect external features. Furthermore, while races differ in allele frequencies, these same alleles are found in most races. In other words, at the genetic level there are no significant discontinuities between races. It is paradoxical that race, which has been so important to people throughout the course of human history, is trivial at the genetic level—an important insight to emerge from genetic analysis.

MENDELIAN INHERITANCE

Gregor Mendel, the Moravian Augustinian monk who founded the modern science of genetics.

Mendelian inheritance is a type of biological inheritance that follows the laws originally proposed by Gregor Mendel in 1865 and 1866 and re-discovered in 1900. These laws were initially controversial. When Mendel's theories were integrated with the Boveri–Sutton chromosome theory of inheritance by Thomas Hunt Morgan in 1915, they became the core of classical genetics. Ronald Fishercombined these ideas with the theory of natural selection in his 1930 book *The Genetical Theory of Natural Selection*, putting evolution onto a mathematicalfooting and forming the basis for population genetics within the modern evolutionary synthesis.

Mendel's Laws

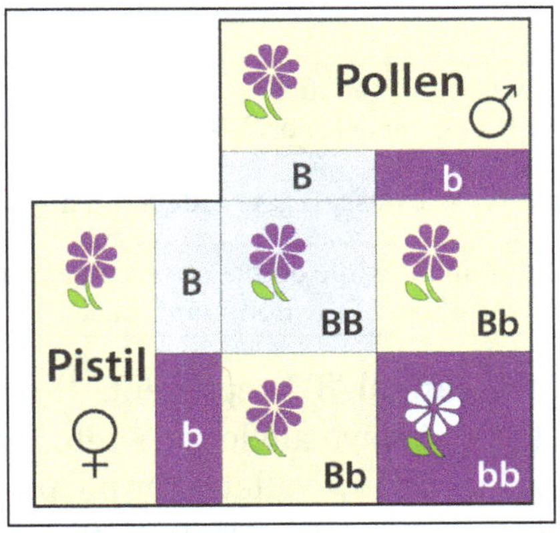

An illustration of Punnett square for one of Mendel's pea plant experiments – self-fertilization of the F1 generation.

Mendel discovered that, when he crossed purebred white flower and purple flower pea plants (the parental or P generation), the result was not a blend. Rather than being a mix of the two, the offspring (known as the F_1 generation) was purple-flowered. When Mendel self-fertilized the F_1 generation pea plants, he obtained a purple flower to white flower ratio in the F_2 generation of 3 to 1. The results of this cross are tabulated in the Punnett square to the right.

He then conceived the idea of heredity units, which he called "factors". Mendel found that there are alternative forms of factors—now called genes—that account for variations in inherited characteristics. For example, the gene for flower color in pea plants exists in two forms, one for purple and the other for white. The alternative "forms" are now called alleles. For each biological trait, an organism inherits two alleles, one from each parent. These alleles may be the same or different. An organism that has two identical alleles for a gene is said to be homozygous for that gene (and is called a homozygote). An organism that has two different alleles for a gene is said be heterozygous for that gene (and is called a heterozygote).

Mendel hypothesized that allele pairs separate randomly, or segregate, from each other during the production of gametes: egg and sperm. Because allele pairs separate during gamete production, a sperm or egg carries only one allele for each inherited trait. When sperm and egg unite at fertilization, each contributes its allele, restoring the paired condition in the offspring. This is called the Law of Segregation. Mendel also found that each pair of alleles segregates independently of the other pairs of alleles during gamete formation. This is known as the Law of Independent Assortment.

The genotype of an individual is made up of the many alleles it possesses. An individual's physical appearance, or phenotype, is determined by its alleles as well as by its environment. The presence of an allele does not mean that the trait will be expressed in the individual that possesses it. If the two alleles of an inherited pair differ (the heterozygous condition), then one determines the organism's appearance and is called the dominant allele; the other has no noticeable effect on the organism's appearance and is called the recessive allele. Thus, in the example above the dominant purple flower allele will hide the phenotypic effects of the recessive white flower allele. This is known as the Law of Dominance but it is not a transmission law: it concerns the expression of the genotype. The upper case letters are used to represent dominant alleles whereas the lowercase letters are used to represent recessive alleles.

Mendel's laws of inheritance	
Law	Definition
Law of segregation	During gamete formation, the alleles for each gene segregate from each other so that each gamete carries only one allele for each gene.
Law of independent assortment	Genes of different traits can segregate independently during the formation of gametes.
Law of dominance	Some alleles are dominant while others are recessive; an organism with at least one dominant allele will display the effect of the dominant allele.

In the pea plant example above, the capital "B" represents the dominant allele for purple flowers and lowercase "b" represents the recessive allele for white flowers. Both parental plants were true-breeding, and one parental variety had two alleles for purple flowers (*BB*) while the other had two alleles for white flowers (*bb*). As a result of fertilization, the F_1 hybrids each inherited one allele for purple flowers and one for white. All the F_1 hybrids (*Bb*) had purple flowers, because the dominant *B* allele has its full effect in the heterozygote, while the recessive *b*allele has no effect on flower color. For the F_2 plants, the ratio of plants with purple flowers to those with white flowers (3:1) is called the phenotypic ratio. The genotypic ratio, as seen in the Punnett square, is 1 *BB* : 2 *Bb* : 1 *bb*.

Law of Segregation of Genes ("First Law")

The Law of Segregation states that every individual organism contains two alleles for each trait, and that these alleles segregate (separate) during meiosis such that each gamete contains only one of the alleles. An offspring thus receives a pair of alleles for a trait by inheriting homologous chromosomes from the parent organisms: one allele for each trait from each parent.

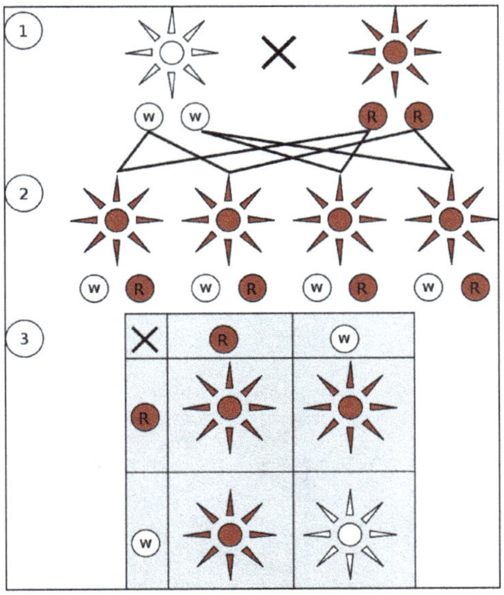

Dominant and recessive phenotypes. (1) Parental generation. (2) F_1 generation. (3) F_2 generation. Dominant (red) and recessive (white) phenotype look alike in the F_1 (first) generation and show a 3:1 ratio in the F_2 (second) generation.

Molecular proof of this principle was subsequently found through observation of meiosis by two scientists independently, the German botanist Oscar Hertwig in 1876, and the Belgian zoologist Edouard Van Beneden in 1883. Paternal and maternal chromosomes get separated in meiosis and the alleles with the traits of a character are segregated into two different gametes. Each parent contributes a single gamete, and thus a single, randomly successful allele copy to their offspring and fertilization.

Law of Independent Assortment ("Second Law")

The Law of Independent Assortment states that alleles for separate traits are passed independently of one another. That is, the biological selection of an allele for one trait has nothing to do with the selection of an allele for any other trait. Mendel found support for this law in his dihybrid cross experiments. In his monohybrid crosses, an idealized 3:1 ratio between dominant and recessive phenotypes resulted. In dihybrid crosses, however, he found a 9:3:3:1 ratios. This shows that each of the two alleles is inherited independently from the other, with a 3:1 phenotypic ratio for each.

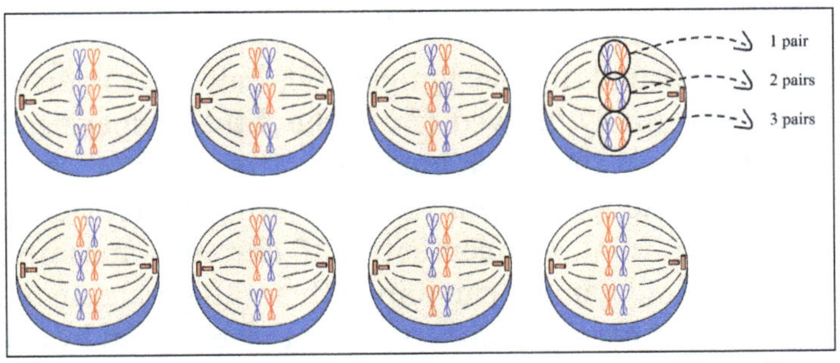

According to independent assortment, 3 homologous pairs create 8 possible combinations, all equally likely to be fertilized. The equation to determine the number of possible combinations given the number of homologous pairs = 2^x (x = number of homologous pairs).

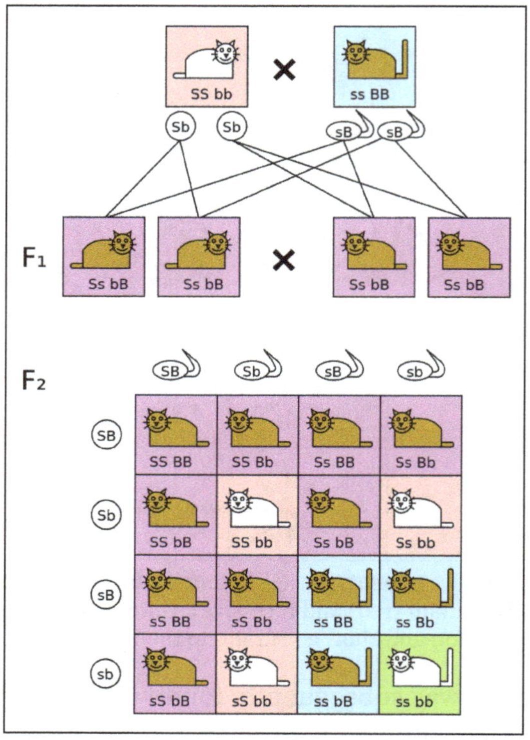

Dihybrid cross: The phenotypes of two independent traits show a 9:3:3:1 ratio in the F_2 generation. In this example, coat color is indicated by B (brown, dominant) or b (white), while tail length is indicated by S (short, dominant) or s (long). When parents are homozygous for each trait (SSbb and ssBB), their children in the F_1 generation are heterozygous at both loci and only show the dominant phenotypes (SsbB). If the children mate with each other, in the F_2generation all combinations of coat color and tail length occur: 9 are brown/short (purple boxes), 3 are white/short (pink boxes), 3 are brown/long (blue boxes) and 1 is white/long (green box).

Independent assortment occurs in eukaryotic organisms during meiotic metaphase I, and produces a gamete with a mixture of the organism's chromosomes. The physical basis of the independent assortment of chromosomes is the random orientation of each bivalent chromosome along the metaphase plate with respect to the other bivalent chromosomes. Along with crossing over, independent assortment increases genetic diversity by producing novel genetic combinations.

There are many violations of independent assortment due to genetic linkage.

Of the 46 chromosomes in a normal diploid human cell, half are maternally derived (from the mother's egg) and half are paternally derived (from the father's sperm). This occurs as sexual reproduction involves the fusion of two haploidgametes (the egg and sperm) to produce a new organism having the full complement of chromosomes. During gametogenesis—the production of new gametes by an adult—the normal complement of 46 chromosomes needs to be halved to 23 to ensure that the resulting haploid gamete can join with another gamete to produce a diploid organism. An error in the number of chromosomes, such as those caused by a diploid gamete joining with a haploid gamete, is termed aneuploidy.

In independent assortment, the chromosomes that result are randomly sorted from all possible maternal and paternal chromosomes. Because zygotes end up with a random mix instead of a pre-defined "set" from either parent, chromosomes are therefore considered assorted independently. As such, the zygote can end up with any combination of paternal or maternal chromosomes. Any of the possible variants of a zygote formed from maternal and paternal chromosomes will occur with equal frequency. For human gametes, with 23 pairs of chromosomes, the number of possibilities is 2^{23} or 8,388,608 possible combinations.The zygote will normally end up with 23 chromosomes pairs, but the origin of any particular chromosome will be randomly selected from paternal or maternal chromosomes. This contributes to the genetic variability of progeny.

Law of Dominance

Mendel's Law of Dominance states that recessive alleles will always be masked by dominant alleles. Therefore, a cross between a homozygous dominant and a homozygous recessive will always express the dominant phenotype, while still having a heterozygous genotype. The Law of Dominance can be explained easily with the help of a mono hybrid cross experiment:- In a cross between two organisms pure for any pair (or pairs) of contrasting traits (characters), the character that appears in the F1 generation is called "dominant" and the one which is suppressed (not expressed) is called "recessive." Each character is controlled by a pair of dissimilar factors. Only one of the characters expresses. The one which expresses in the F1 generation is called Dominant. However, the law of dominance is not universally applicable.

Mendelian Trait

A Mendelian trait is one that is controlled by a single locus in an inheritance pattern. In such cases, a mutation in a single gene can cause a disease that is inherited according to Mendel's laws. Examples include sickle-cell anemia, Tay–Sachs disease, cystic fibrosis and xeroderma pigmentosa. A disease controlled by a single gene contrasts with a multi-factorial disease, like heart disease, which is affected by several loci (and the environment) as well as those diseases inherited in a non-Mendelian fashion.

Non-Mendelian Inheritance

Mendel explained inheritance in terms of discrete factors—genes—that are passed along from generation to generation according to the rules of probability. Mendel's laws are valid for all sexually reproducing organisms, including garden peas and human beings. However, Mendel's laws stop short of explaining some patterns of genetic inheritance. For most sexually reproducing organisms, cases where Mendel's laws can strictly account for the patterns of inheritance are relatively rare. Often, the inheritance patterns are more complex.

The F_1 offspring of Mendel's pea crosses always looked like one of the two parental varieties. In this situation of "complete dominance," the dominant allele had the same phenotypic effect whether present in one or two copies. But for some characteristics, the F_1 hybrids have an appearance *in between* the phenotypes of the two parental varieties. A cross between two four o'clock (*Mirabilis jalapa*) plants shows this common exception to Mendel's principles. Some alleles are neither dominant nor recessive. The F_1 generation produced by a cross between red-flowered (RR) and white flowered (WW) *Mirabilis jalapa* plants consists of pink-colored flowers. Neither allele is dominant in this case. This third phenotype results from flowers of the heterzygote having less red pigment than the red homozygotes. Cases in which one allele is not completely dominant over another are called incomplete dominance. In incomplete dominance, the heterozygous phenotype lies somewhere between the two homozygous phenotypes.

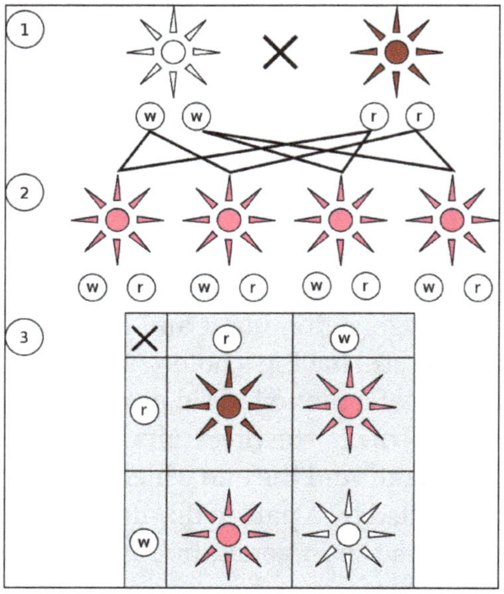

In four o'clock plants, the alleles for red and white flowers show incomplete dominance. As seen in the F_1generation, heterozygous (*wr*) plants have "pink" flowers—a mix of "red" (rr) and "white" (ww) coloring. The F_2generation shows a 1:2:1 ratio of red:pink:white.

A similar situation arises from codominance, in which the phenotypes produced by both alleles are clearly expressed. For example, in certain varieties of chicken, the allele for black feathers is codominant with the allele for white feathers. Heterozygous chickens have a color described as "erminette", speckled with black and white feathers. Unlike the blending of red and white colors in heterozygous four o'clocks, black and white colors appear separately in chickens. Many human genes, including one for a protein that controls cholesterol levels in the blood, show codominance, too. People with the heterozygous form of this gene produce two different forms of the protein, each with a different effect on cholesterol levels.

In Mendelian inheritance, genes have only two alleles, such as *a* and *A*. In nature, such genes exist in several different forms and are therefore said to have multiple alleles. A gene with more than two alleles is said to have multiple alleles. An individual, of course, usually has only two copies of each gene, but many different alleles are often found within a population. One of the best-known examples is coat color in rabbits. A rabbit's coat color is determined by a single gene that has at least four different alleles. The four known alleles display a pattern of simple dominance that can produce four coat colors. Many other genes have multiple alleles, including the human genes for ABO blood type.

Furthermore, many traits are produced by the interaction of several genes. Traits controlled by two or more genes are said to be polygenic traits. *Polygenic* means "many genes." For example, at least three genes are involved in making the reddish-brown pigment in the eyes of fruit flies. Polygenic traits often show a wide range of phenotypes. The broad variety of skin color in humans comes about partly because at least four different genes probably control this trait.

EVOLUTION

In biology, evolution is the change in the characteristics of a species over several generations and relies on the process of natural selection.

- The theory of evolution is based on the idea that all species? are related and gradually change over time.
- Evolution relies on there being genetic variation? in a population which affects the physical characteristics (phenotype) of an organism.
- Some of these characteristics may give the individual an advantage over other individuals which they can then pass on to their offspring.

Natural Selection

- Charles Darwin's theory of evolution states that evolution happens by natural selection.
- Individuals in a species show variation in physical characteristics. This variation is because of differences in their genes.
- Individuals with characteristics best suited to their environment are more likely to survive, finding food, avoiding predators and resisting disease. These individuals are more likely to reproduce and pass their genes on to their children.
- Individuals that are poorly adapted to their environment are less likely to survive and reproduce. Therefore their genes are less likely to be passed on to the next generation.
- As a consequence those individuals most suited to their environment survive and, given enough time, the species will gradually evolve.

Natural Selection in Action: The Peppered Moth

- Before the industrial revolution in the mid-1700s, the peppered moth was most commonly a pale whitish colour with black spots.
- This colouring enabled them to hide from potential predators on trees with pale-coloured bark, such as birch trees.
- The rarer dark-coloured peppered moths were easily seen against the pale bark of trees and therefore more easily seen by predators.
- As the Industrial Revolution reached its peak, the air in industrial areas became full of soot. This stained trees and buildings black.
- As a result, the lighter moths became much easier to spot than the darker ones, making them vulnerable to being eaten by birds.
- The darker moths were now camouflaged against the soot-stained trees and therefore less likely to be eaten.

- Over time this change in the environment led to the darker moths becoming more common and the pale moths rarer.

A pale peppered moth on an oak tree.

A pale peppered moth on a dark tree.

What have Genes got to do with it?

- The mechanisms of evolution operate at the genomic level. Changes in DNA? sequences affect the composition and expression? of our genes, the basic units of inheritance?.
- To understand how different species have evolved we have to look at the DNA sequences in their genomes.
- Our evolutionary history is written into our genome. The human genome looks the way it does because of all the genetic changes that affected our ancestors.
- When DNA and genes in different species look very similar, this is usually taken as evidence of them sharing ancestors.
- For example, humans and the fruit fly, Drosophila melanogaster, share much of their DNA. 75 per cent of genes that cause diseases in humans are also found in the fruit fly.
- DNA accumulates changes over time. Some of these changes can be beneficial, and provide a selective advantage for an organism.
- Other changes may be harmful if they affect an important, everyday function. As a result some genes do not change much. They are said to be conserved.

COMMON DESCENT

Common descent is a term within evolutionary biology which refers to the common ancestry of a particular group of organisms. The process of common decent involves the formation of new species from an ancestral population. When a recent common ancestor is shared between two organisms, they are said to be closely related. In contrast, common descent can also be traced back to a universal common ancestor of all living organisms using molecular genetic methods. Such evolution from a universal common ancestor is thought to have involved several speciation events as a result of natural selection and other processes, such as geographical separation.

Theory of Common Descent

The theory of common descent states that all living organisms are descendants of a single ancestor. Thus, the Theory of Common Descent helps to explain why species living in different geographical regions exhibit different traits, some traits are highly conserved among broad animal classifications (e.g., vertebrates or tetrapods), seemly different species (e.g., birds and reptiles) share inherited physical and genetic traits, and successfully adapted organisms typically produce more offspring. While the Theory of Common Descent is primarily derived from the physical observation of various phenotypes (e.g., size, colour, beak shape, embryological development, etc.), modern advances in genetics and associated molecular techniques have been able to demonstrate that the process by which DNA is eventually translated into proteins is maintained among all lifeforms. Small changes in DNA between organisms have revealed a shared ancestry as well as insight into important changes that resulted in various speciation events. Phylogenetic trees and cladograms are often used to hypothesize the evolution of various organisms and shared common descent.

Examples of Common Descent

Human Chromosome 2

Compelling evidence of the shared common ancestry of humans with the great apes is the fusion event which occurred when two chromosomes common in apes fused to form chromosome 2 in humans. This resulted in humans having only 23 pairs of chromosomes, while all other hominidae have 24 pairs. The great apes (e.g., chimpanzees, gorillas, and orangutans) have two chromosomes with almost identical DNA sequences as that found in chromosome 2 of humans. Further evidence of such a fusion event is the residual presence of telomeres and a centromere, which indicate that the genetic information was historically found on two separate chromosomes.

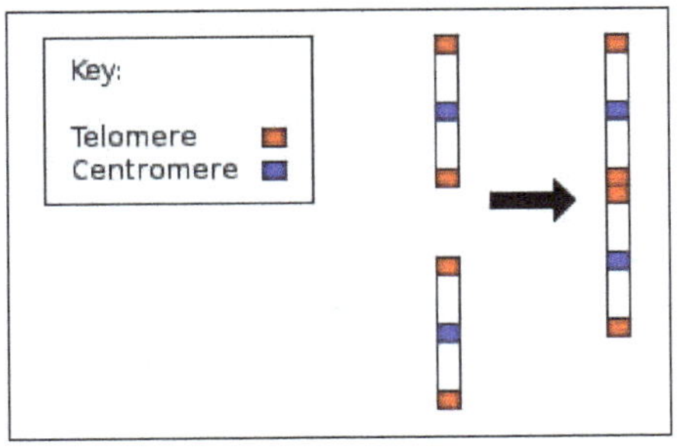

Endogenous Retroviruses

Endogenous retroviruses are residual DNA sequences found in the genomes of virtually all living organisms as a result of ancient viral infections. Since the retroviral sequences are incorporated into the DNA of the host organism, such sequences are inherited in the offspring. Since such infections are random events, as is the location in which the viral genome is inserted within the genome, the identification of the same retroviral sequences in multiple species is indicative of a shared ancestry. Such analyses of endogenous retroviruses often reveal speciation events (e.g., feline endogenous retroviruses reveal the separation between large and small cat species) and how closely related two species may be, as observed in the shared endogenous retroviruses between humans and other primate species.

Presence of Atavisms

Atavisms are the appearance of a lost trait observed in an ancestral species that is not observed in more recent ancestors. Atavisms are an example of common descent as they provide evidence of the phenotypical or vestigial features that are often retained throughout evolution. Some examples, include the appearance of hind limbs in whales as evidence of a terrestrial ancestor, teeth exhibited by chickens, additional toes observed in modern horse species, and the back flippers of bottlenose dolphins. Atavisms tend to arise because the ancestral genes are not deleted from the genome, but rather silenced and then reactivated in later offspring.

Vestigial Structures

Similar to atavisms, vestigial structures are structures that are homologous to those found in ancestral species, but have become underdeveloped, non-functional, or degenerated in more recently evolved organisms. Such structures provide evidence of adaptations to a new environment, in which the ancestral organ or limb is no longer required, or has been modified to better suit a new purpose. There are an abundance of examples of vestigial structures observed in nature. Some examples include, the hind limbs and pelvic girdle observed in whales (as shown below) and snakes, non-functional wings exhibited by some insect species, non-functional wings of flightless bird species (e.g., ostriches), abdominal segments of barnacles, and the embryonic limb buds exhibited by several species lacking hind limbs (e.g., dolphins).

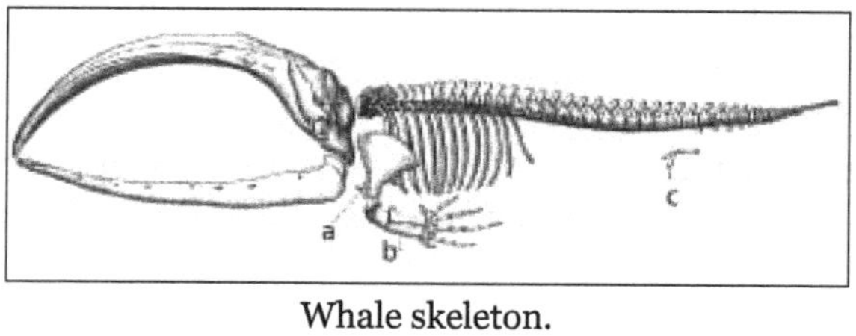

Whale skeleton.

Pentadactyl Limbs

The presence of pentadactyl limbs is an example of a homologous trait exhibited by all tetrapods and is highly conserved throughout evolution, despite some modifications. Such limbs are first observed in the evolution from fishes to amphibians and consists of a single proximal, two distal, carpal, five metacarpal, and phalange bones. Although the overarching structure of the pentadactyl limbs is similar, various modifications have been made throughout evolution as adaptations to specific environments or lifestyles. Some examples include the modified pentadactyl wings of bats, the elongated forearms of primate species, the flippers of dolphin and whale species, and the modified digits of horses to form a hoof.

Fossil Evidence

The fossilized remains of various organisms combined with modern dating methods provides some of the most convincing evidence of common descent and evolutionary history. Fossilization occurs when the bones of a decaying animal become porous and the mineral salts in the surrounding earth infiltrate the bones, converting them to stone. Other methods include preservation in ice, imprinted remains (e.g., plants or footprints), tree resin, and peat. Since fossils are found in sedimentary rock, which is formed by layers of silt and mud, each layer corresponds to a specific geological period which can be dated. Thus, the extinction, evolution, and emergence of various species can be observed throughout history using the fossil record. Many extinct species are also observed in the fossil record, such as dinosaurs.

Biogeography

Biogeography presents compelling evidence of common descent by showcasing speciation and novel traits through adaptations to environmental pressures. One of the most famous examples is that of island biogeography and Charles Darwin's observations of the beaks of finches residing on the Galapagos Islands. In these finches, the beaks had been adapted for the specific vegetation found on the island, resulting in deviation from the ancestral finches found on the mainland. Long term effects of geographical separation are also observed with the evolution of new species that are not found elsewhere in the world. An example of this is the presence of marsupial species on the continent of Australia, and the emergence of polar bears as a result of geographical isolation in the arctic.

SPECIATION

Speciation is a process within evolution that leads to the formation of new, distinct species that are reproductively isolated from one another.

Anagenesis, or 'phyletic evolution', occurs when evolution acts to create new species, which are distinct from their ancestors, along a single lineage, through gradual changes in physical or genetic traits. In this instance, there is no split in the phylogenetic tree. Conversely, 'speciation' or cladogenesis arises from a splitting event, where a parent species is split into two distinct species, often as the result of geographic isolation or another driving force involving the separation of populations.

The reproductive isolation that is integral to the process of speciation occurs due to reproductive barriers, which are formed as a consequence of genetic, behavioral or physical differences arising between the new species. These are either pre-zygotic (pre-mating) mechanisms, for example, differences in courtship rituals, non-compatible genitalia, or gametes, which are unable to fertilize between species. Alternatively, they are post-zygotic (post-mating), for example zygote mortality or the production of sterile offspring. Reproductive isolation leads to reinforcement of the distinction between species through natural selection and sexual selection.

Types of Speciation

Allopatric Speciation

Allopatric speciation occurs when members of a population become geographically isolated from one another, to the extent that genetic exchange, through mating, is prevented or interfered with. This may be a result of geographical changes, such as the formation of a mountain by a volcano, island formation, habitat separation by glaciers and rivers, or habitat fragmentation caused by human activity. Alternatively, species members may emigrate, resulting in population separation by dispersal; this is commonly known as vicariance.

The separated populations then undergo divergence in genotypic or phenotypic traits as a result of different selective pressures acting upon populations. This leads natural selection to cause genetic drift as mutations arise within populations. Over time, the separate populations may develop morphologically distinct features due to adaption to their new environment. The features may become so distinctively different that reproductive isolation occurs, preventing the inbreeding of populations and thus forming new species. If the populations become sufficiently different that they are classified as new species, but not distinct enough for reproductive isolation to occur, the species may come back into contact and mate, producing hybrids.

The extent of the effect that geographic barriers may have on a population often depends on the dispersal ability of the organism; for example, the new formation of a river in a landscape would create an impassable barrier for small terrestrial mammals, insects and reptiles. However, birds and larger mammals would likely disperse across the river with ease.

An elegant example of allopatric speciation, which first inspired Charles Darwin to develop the theory of evolution and natural selection, is the divergent populations of finches inhabiting the Galapagos Islands, and known as 'Darwin's finches'. Darwin noticed that each of the Galapagos Islands hosted a population of finches, which although relatively similar in morphology (compared with other bird species), exhibited slight differences in features such as body size, color and beak length or shape. He noted that there were different food sources available for the birds on each of the different islands, and came to the conclusion that the differences in beak shape were an adaption toward acquiring the particular food source.

Evolution of Finches: 1. Geospiza maggnirostris, 2. Geospiza fortis, 3. Geospiza parvula, 4. certhidea olivasea.

Sympatric Speciation

Sympatric speciation is the evolutionary process whereby species are formed from a single ancestral species while inhabiting the same geographic area. In contrast to allopatric speciation, the distribution ranges of species which evolve through sympatry may be identical or they may only overlap. Rather than geographic distance prompting a reduction of gene flow between populations, sympatry occurs when members of one population make use of a new niche. This could occur, for example, if a herbivorous insect begins to feed on a new or novel plant source with which it was not ancestrally associated, or if a new species of plant is introduced to the species' geographic range. As insects generally reproduce or lay eggs within the type of fruit that they were born in, over time, the individuals would specialize in feeding and mating on particular fruits. Consequently, gene flow between populations that specialize in different fruits will be reduced, leading to reproductive isolation of the populations. It is possible that the populations will also develop morphological differences as they adapt to most effectively exploit the new niche. Although sympatric speciation does sometimes occur, it is uncommon, especially within large, multicellular organisms.

Parapatric Speciation

Parapatric speciation is an extremely rare case of speciation that occurs when a population is continuously distributed within a geographic area without any specific barriers to gene flow. Nonetheless, the population does not mate randomly within the population, but rather individuals mate more commonly with their closest geographic neighbors, resulting in uneven gene flow. Non-random mating may increase the rate of dimorphism within populations, in which varied morphological forms of the same species are displayed. The result of parapatric speciation is one or more distinct sub-populations (known as 'sister species'), which have small, continuous overlaps in their biogeographic range and are genotypically dimorphic.

Peripatric Speciation

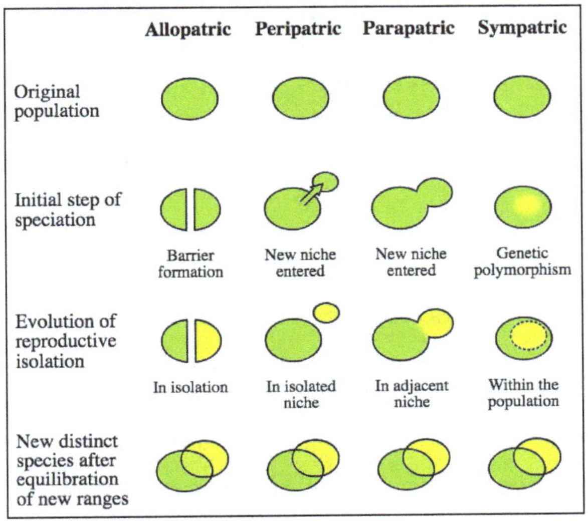

Speciation modes.

Peripatric speciation is a form of allopatric speciation that occurs when populations that have become isolated have very few individuals. Through this process, the population goes through a genetic bottleneck. Within the small sub-population, organisms which are able to survive within the new environment may carry genes that were rare within the main population but that cause a slight variation to behavior or morphology. Through repeated matings, the frequency of these, once rare, genes increases within the small population. This is known as the 'founder effect'. Over time, the characteristic that was determined by the gene becomes fixed within the population, leading to an isolated species that is evolutionarily distinct from the main population.

Artificial Speciation

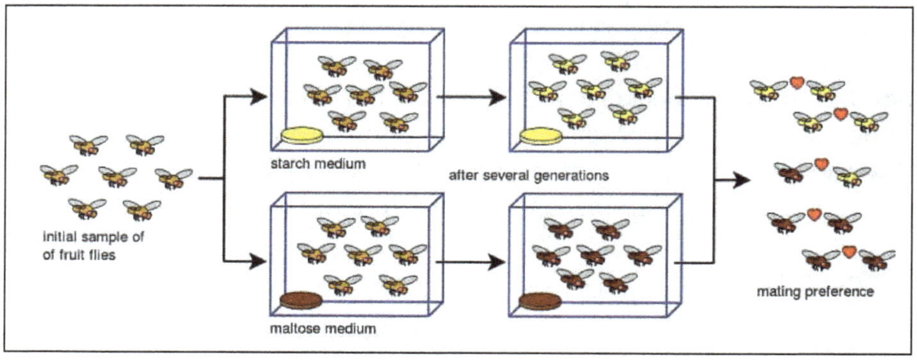

Drosophila speciation experiment.

Artificial speciation is the form of speciation that can be achieved by the input of human influence. By separating populations, and thereby preventing breeding, or by intentionally breeding individuals with desired morphological or genotypic traits, humans can create new, distinct species. This is also known as 'artificial selection'; most modern domesticated animals and plants have undergone artificial selection.

Although evolution of our modern crops and livestock has taken thousands of years, it is possible to visualize the process of artificial selection in species that have short life cycles. Artificial selection has been demonstrated most effectively in species of Fruit Fly (Drosophila melanogaster). Experiments in which flies are placed into environments which contain different resources or habitats show the changes that occur when the flies adapt to each environment. After several generations, the flies are removed from the experimental zone and are allowed to cohabitate, although the populations are unable to mate due to the reproductive isolation process that occurred while in isolation.

References

- Heredity-genetics, science: britannica.com, Retrieved 29 June, 2019
- Grafen, Alan; Ridley, Mark (2006). Richard Dawkins: How A Scientist Changed the Way We Think. New York, New York: Oxford University Press. P. 69. ISBN 978-0-19-929116-8
- What-is-evolution, facts: yourgenome.org, Retrieved 30 January, 2019
- Bailey, Regina (5 November 2015). "Mendel's Law of Segregation". About education. About.com. Retrieved 2 February 2016
- Common-descent: biologydictionary.net, Retrieved 31 March, 2019
- Speciation: biologydictionary.net, Retrieved 14 July, 2019

PERMISSIONS

All chapters in this book are published with permission under the Creative Commons Attribution Share Alike License or equivalent. Every chapter published in this book has been scrutinized by our experts. Their significance has been extensively debated. The topics covered herein carry significant information for a comprehensive understanding. They may even be implemented as practical applications or may be referred to as a beginning point for further studies.

We would like to thank the editorial team for lending their expertise to make the book truly unique. They have played a crucial role in the development of this book. Without their invaluable contributions this book wouldn't have been possible. They have made vital efforts to compile up to date information on the varied aspects of this subject to make this book a valuable addition to the collection of many professionals and students.

This book was conceptualized with the vision of imparting up-to-date and integrated information in this field. To ensure the same, a matchless editorial board was set up. Every individual on the board went through rigorous rounds of assessment to prove their worth. After which they invested a large part of their time researching and compiling the most relevant data for our readers.

The editorial board has been involved in producing this book since its inception. They have spent rigorous hours researching and exploring the diverse topics which have resulted in the successful publishing of this book. They have passed on their knowledge of decades through this book. To expedite this challenging task, the publisher supported the team at every step. A small team of assistant editors was also appointed to further simplify the editing procedure and attain best results for the readers.

Apart from the editorial board, the designing team has also invested a significant amount of their time in understanding the subject and creating the most relevant covers. They scrutinized every image to scout for the most suitable representation of the subject and create an appropriate cover for the book.

The publishing team has been an ardent support to the editorial, designing and production team. Their endless efforts to recruit the best for this project, has resulted in the accomplishment of this book. They are a veteran in the field of academics and their pool of knowledge is as vast as their experience in printing. Their expertise and guidance has proved useful at every step. Their uncompromising quality standards have made this book an exceptional effort. Their encouragement from time to time has been an inspiration for everyone.

The publisher and the editorial board hope that this book will prove to be a valuable piece of knowledge for students, practitioners and scholars across the globe.

INDEX

www.ingramcontent.com/pod-product-compliance
Lightning Source LLC
LaVergne TN
LVHW080711210926
848485LV00003B/447